THE SEAS AND OCEANS IN COLOUR

THE SEAS AND OCEANS

in Colour

C. F. Hickling
and
Peter Lancaster Brown

BLANDFORD PRESS
LONDON

First published 1973

167 High Holborn,
London WC1V 6PH

ISBN 0 7137 0646 5

Colour section printed by Colour Reproductions Ltd, Billericay, Essex
Set in Photon Imprint 10 pt. by
Richard Clay (The Chaucer Press) Ltd, Bungay, Suffolk
and printed in Great Britain by
Fletcher & Son Ltd, Norwich

CONTENTS

		Page
ACKNOWLEDGMENTS		vi
INTRODUCTION		viii

Part 1: The Nature of the Seas and Oceans

1	ORIGIN OF THE HYDROSPHERE	1
2	DISTRIBUTION, COMPOSITION AND SOME PROPERTIES OF SEA-WATER	6
3	CONFIGURATIONS OF THE SEAS AND OCEANS	11
4	MOVEMENT OF SEA-WATER	22
5	EVOLUTION OF THE OCEANS AND CONTINENTAL DRIFT	48
6	THE ENVIRONMENTS	64
7	ISLANDS AND CORAL REEFS	74

Part 2: Man Explores the Seas and Oceans

8	HISTORY OF MARINE EXPLORATION AND OCEANOGRAPHY	167
9	TECHNOLOGY AND THE OCEANS	178
10	DESALINATION	194
11	POWER FROM THE SEA	198
12	BASIC FOOD RESOURCES OF THE OCEANS	202
13	FISHING AND FISH FARMING	214
14	CONSERVATION OF THE RESOURCES OF THE SEA	226
15	WHO OWNS THE OCEANS AND SEABED?	237
INDEX		240

ACKNOWLEDGMENTS

The authors and publishers are grateful to those listed below for permission to reproduce illustrations.

Cover illustration, British Gas Corporation; endpapers, Roland Scherman; Fig. 16, U.S. Navy; Fig. 17, The National Maritime Museum, London; Fig. 18, by kind permission of the President and Council of the Royal College of Surgeons of England; Fig. 19, National Portrait Gallery, London; Fig. 30, *inset*, copyright Solarfilma, Reykjavik, Iceland; Figs 37 and 38, Australian Information Service; Fig. 47, *insets*, *potash*, Israel Government Tourist Office, *gas rig*, British Gas Corporation, *salt farming*, Australian Information Service; Fig. 61, *insets*, G. R. Williamson, Dolphin Research Sta Station, Kessock, Inverness; Fig. 66, *upper inset*, French Information Service; Fig. 72, Picturepoint Ltd; Fig. 73, Norman Myers/Bruce Coleman Ltd; Fig. 71, *inset*, The Weir Group Ltd; Fig. 74, Shell International Petroleum Company Limited; Fig. 75, Dr Geoffrey Potts, Marine Biological Association; Figs 76 and 77, Dr Molly Spooner, Marine Biological Association; Fig. 79, Dr Joseph R. Jehl Jr, Natural History Museum, San Diego, California; Fig. 80, *insets*, *side trawler*, White Fish Authority, *Dutch beam-trawler*, *handling trawl*, *big catch on deck*, Ministry of Agriculture, Fisheries and Food; Fig. 81, *Scottish fishing craft*, White Fish Authority; Fig. 82, *insets*, White Fish Authority; Fig. 83, Australian Information Service; Fig. 84, *both pictures*, Dr N. R. Merrett, National Institute of Oceanography; Fig. 85, Anne Bolt; Fig. 86, Japan Information Centre; Fig. 87, *both pictures*, Ministry of Agriculture, Fisheries and Food; Fig. 88, *both pictures*, Graham Smart, Joseph Johnston & Sons Ltd, Montrose; Fig. 89, *Philippines fish farming*, courtesy of the U.N. Food and Agriculture Organization, *Japanese prawn farm*, Japan Information Centre.

Acknowledgment is also given for Fig. 2, modified after Fisher and Revelle 1955; Fig. 3, modified after P. A. Gordienko 1961; Fig. 8, modified after J. Berstein 1954; Fig. 9, after Antonio Snider-Pellegrini 1858; Fig. 12, modified after H. W. Menard 1969; Fig. 14, modified after J. A. Steers; Fig. 35, modified after J. P. Isaacs 1969; Fig. 56, modified after E. Steeman-Nielsen, *Galathea* reports; Fig. 58, modified after F. S. Russell and C. M. Yonge; Fig. 62, modified after Murray and Hjort, *Galathea* reports; Fig. 63, modified after Olga Marshall and Murray and Hjort; Fig. 91, modified after F. S. Russell and C. M. Yonge. Other photographs and sources were supplied by the authors, and all coloured illustrations are by John W. Wood. Black and white text drawings are by Peter Lancaster Brown.

We would also like to thank the following for assistance with picture references: The Grumman Aircraft Engineering Corporation (Fig. 27, *Ben Franklin*), James Hall (Photographers), Greenock, and the Natural Environment Research Council (Fig. 33), the U.S. Navy (Fig. 40), and Novosti Press Agency (Fig. 65).

The authors would also like opportunity to thank the staff of Blandford Press for their general enthusiasm and helpfulness in all matters; and Peter Lancaster Brown wishes to acknowledge the generous assistance from his wife, Johanne, who shared in much of the routine drudgery.

INTRODUCTION

The seas and oceans have always held a fascination for mankind, and in modern times we are sometimes inclined to forget that ancient man had intimate knowledge of the sea routes of the world. Much of this early knowledge was lost for, apart from intriguing snippets of evidence, most of their voyages are not recorded, and extant documents relating to almost every voyage prior to about A.D. 1000 are extremely rare.

Some of the ports from which early man set sail are no longer at the sea edge, such as the great delta ports in the Persian Gulf, which reached their peak in about 4000 B.C. Recently pre-historic artefacts have been traced from these ports across the ancient ocean trade routes which extended to the far corners of the Earth, and this evidence is fully supported by radiocarbon datings. The later mediaeval voyagers referred to sailing 'the seven seas', and the term was reintroduced into romantic literature about the sea at the end of the nineteenth century when Rudyard Kipling used it as a title for a book of his poems.*

The first scientific ideas about the seas were formulated by the Greeks, and several of these remain the basis of present-day ideas. During the past decade, ocean science in particular has had considerable influence in changing some of the long-established concepts about the Earth, and the revolutionary phenomenon of sea-floor spread has been described as the trinity of the Earth sciences. Sea or ocean science is nowadays called *oceanography*, or by an even newer word, *oceanology*, coined in the 1960s. Some like to differentiate between the use of these terms, using *oceanography* to describe the more academic aspects of ocean science, and *oceanology* to describe modern ocean technology, including the political, legal, financial and

* The ancient 'seven seas' were the Mediterranean, Red, China, West Africa and East Africa seas, the Indian Ocean and the Persian Gulf. The modern unofficial 'seven seas' comprise the Antarctic, Arctic, North Atlantic, South Atlantic, North Pacific, South Pacific and Indian oceans.

economic aspects of the seas and oceans. Sir John Murray, one of the principal scientists of the H.M.S. *Challenger* Expedition (1872–6), believed he had invented the word *oceanography* in about 1880, but he later discovered that the word *oceanographie* was first used in France about 1584, but then dropped out of usage.*

In popular literature it has become a journalistic cliché to refer to the enormous untapped and inexhaustible resources of the oceans, and the great role they will surely play in man's life in the future. As a generality it is a cliché which is almost true. The part which is not true, regrettably, is that the resources of the oceans are inexhaustible (see p. 227). Common greed has already decimated vast populations of marine life to the point of extinction. Many of the world's fishing grounds have been overfished and transformed into underwater Saharas. The great whales – the largest animals ever to live on the Earth – have been ruthlessly slaughtered so that the once-important commercial fishery is now uneconomic to pursue. Pollution is manifest on a global scale. Even fish and seals in remote Antarctic waters show dangerously high concentrations of man-made toxic wastes and, nearer to land, local pollution disasters are commonplace occurrences. When Thor Heyerdahl sailed his papyrus boat *Ra* across the Atlantic, he found the sea covered with lumps of coagulated crude-oil spillage.

The seas and oceans have been colloquially called inner space as opposed to outer (cosmic) space. Man's economic and social future hinges on inner space far more than on the hostile environment of outer space. Now that the technological challenge of reaching the Moon and Mars has been realized, scientific technology is now being rechannelled into the hydrosphere. Economic and social pressures have already made many nations look hurriedly towards the oceans. By the year 2000, at the present rate of population growth, most of the water drunk by the Earth's inhabitants will come from desalinated sea-water. Already chemical and mineral exploitation of sea-water is big business. Most of the world's magnesium and bromine come from the seas, and the harvesting of common sea-salt is a fast growing industry in many countries. But it is in the realms of hydrocarbon fossil fuels that the richest commercial rewards are

* The term *ocean engineering* has also been in use in recent times. It has been defined as the ability to do useful work on, in and under the seas and oceans.

at present being reaped. By 1978 over 40 per cent of the world's oil and gas will originate from beneath the ocean bed. Awaiting exploitation in the ever hungrier future is a whole range of valuable economic and strategic minerals such as manganese, diamonds, iron ore, tin, etc.

Yet, perhaps the most exciting scientific and technological challenge is to find ways of exploiting the various basic resources of marine life at present untapped by man's economy. The great whales have almost gone, but the vast stocks of protein-rich krill, on which they feed, now remain only a partially exploited food source in the oceans' food chain. So far man's invention and technology has been unable to provide economic methods of exploiting this vast resource.*

It is not generally realized that the oceans are capable of much greater food production area for area than the land. Whereas on land one hectare (2½ acres) could yield 340 kilograms (750 pounds) of beef, the same oceanic area will produce 2½ tons of fish, or 125 to 250 tons of shell-fish under conditions of intensive cultivation. In the ocean husbandry of the food resources is more important than any other single consideration, but it can only be tackled successfully on an international and global basis so that individual nationalistic greeds, which cause overfishing and pollution, may no longer jeopardize its yet unrealized potential as the protein bank for the future.

P.L.B.

* Although the Soviet Union at present harvests krill (see p. 228), it is doubtful whether their methods are economic in the sense of the criterion as applied in Western fisheries.

PART 1: THE NATURE OF THE SEAS AND OCEANS

1 ORIGIN OF THE HYDROSPHERE *

Earliest Ideas

During the seventeenth century the Irish churchman Archbishop James Ussher declared with complete confidence that the Earth, including the oceans, was created at 9 a.m. on 23 October 4004 B.C. But long before the Archbishop's pronouncement, men had pondered deeply about the origin of the Earth, the oceans and life generally. In world mythologies one of the most striking similarities is the creation myth where mankind arose out of a primeval abyss or waste of water. This legendary watery creation can be observed in races as far apart in time and geography as the Egyptians, Jews, Hindus, Celts and North American Indians. Herodotus said that it was Thales who first proposed that everything was made of water. Anaximenes thought that the oceans were older than the land masses; fish were the first form of life which appeared on Earth, and when dry land rose out of the primeval waters, some of the fish adapted themselves to living on land; hence man evolved from fish.

According to the Greek poet Hesiod (eighth century B.C.), the first to come upon the vacant Earth were children born of the union of Earth and Heaven, and among them was Oceanus, the deep swirling ocean river that encircled the Earth. Aristotle posed the question: Could the sea have come from one source? He was puzzled about the origin of the Earth's water. He knew about the rivers as a source of water – the Nile in particular – and the underground water which supplied the wells. Intuitively he guessed that the seas were different, since they ebbed and flowed, although actually Aristotle had no real understanding of tides.

Aristotle also wondered whether some day in the distant future the seas might dry up. Other Greek philosophers pondered whether the seas were the sweat of the Earth, and in doing so came close to the modern ideas on the subject, but Aristotle thought this notion

* The collective name for all the water of the Earth including that which forms the seas, oceans, lakes, ice-caps and water vapour in the atmosphere.

was absurd. Plato, attempting to analyse this idea logically, said that if the sea ran out of the depths of the Earth, the water would have to perform the proverbial impossibility of flowing uphill.

But to the Greeks the sea was considered the terminal and not the source of the Earth's waters, because all rivers flowed into the sea. However, there remained the fundamental question: If the sea was the terminus, and all the immense rivers poured their water into it, why did it not become progressively larger? Aristotle's explanation was that this was due to evaporation, a phenomenon he understood well: he knew the Sun drew up water to the upper regions, where it condensed and later fell again as rain.

But perhaps the greatest puzzle to the Greeks was the saltiness of the seas. Although it was attributed to an admixture with 'salty earth', why then were the rivers not also salty? And how could the effect be limited to a vast body such as the oceans and not extend to the inland seas? Although the Greeks often came up with the wrong answers, they posed the right questions, which is a fundamental part of the processes of science. It was the framing of questions about the occurrence and origin of natural phenomena that led to the Greeks providing the groundwork of most modern scientific thinking.

Modern Ideas

Although our general knowledge about the nature of the Earth and its oceans has increased enormously since Greek times and the seventeenth century when Archbishop Ussher made his pronouncement, theories about the earliest years of Earth history are still only very speculative ones.

To trace the early history of the Earth we must first examine ideas about how the solar system and the Sun itself came into being before the Earth became a separate cosmic body. The Sun was formed as a result of the coalescence of a cloud of cosmic dust and gas. Most of the gas consisted of hydrogen and helium. At the present time we can witness similar star births taking place in various parts of our own galaxy and, perhaps ironically, we know a good deal more about the various processes involved in star formation than we do about the formation of cool bodies such as the Earth.

Speculation really begins with the ideas describing how the Sun became the central parent body to the host of planets plus all the other multifarious constituents like comets, meteorites, etc., that now belong (at least gravitationally) to the solar family. What seems likely is that the planets were formed from the condensation of a similar cosmic dust cloud to that which formed the Sun. Much of the material of this dust cloud was matter created by the first stars that formed the galaxy. Inside their thermonuclear-furnace interiors, these early stars synthesized all the ninety-two chemical elements which now exist. Subsequently, when these early stars died or exploded, the 'finished' elements were scattered throughout space. Then at a later date they provided material from which new stars and planets evolved.*

Still in the realms of speculation is the mystery surrounding the early conditions on the primeval Earth. Was it a hot or a cold body? During the nineteenth century all theories about the origin of the Earth considered that it was born in a fiery state; then through the immensity of time it slowly cooled to become the world we know to-day. Nowadays, however, an opposite view is held. Many consider that the Earth came about by the gravitational contraction of a cold cloud of dust and gas particles, or was the result of a gradual accretion of a number of different-sized, cold, smaller bodies called planetesimals (large chunks like meteorites) which had solidified during the earlier history of the solar system. But neither hot nor cold theories of the origins of the Earth can account for the present-day composition of the Earth's atmosphere or the oceans. It seems that the atmosphere and the oceans are of secondary origin. Either the Earth originated without an atmosphere, or it lost it during a later thermal episode.

We do know that the age of the Earth is about 4·6 aeons.† The oldest known rocks are 4·0 aeons so that we have a missing 600 million years in the Earth's history. However, the continuous existence of sedimentary rocks dating back for more than 3,000 million years demonstrates a continuity of the atmosphere and hydrosphere over a period of at least that length of time, for they otherwise could not have formed as they did.

The origin of the present-day atmosphere and hydrosphere began with a melting of the crustal rocks of the Earth which occurred *after* the missing 600 million years. Such a melting of the rocks would

* See Peter Lancaster Brown, *Astronomy In Colour* (Blandford Press, 1972) for more detailed explanation of the origin of the stars and the Earth as a planet.

† 1 aeon = one thousand million years.

release the volatiles locked away since the Earth's formation. The source of heating can be accounted for either by radiogenic heating or tidal friction (when gravitational energy would be converted into heat). If the Earth had captured the Moon about this time, this particular event could well explain the source of tidal heat.

One clue to the date of the possible capture of the Moon by the Earth – bringing about the event of tides – are fossil stromatolites, which are dome-shaped sedimentary structures of algal origin found on beaches only between tidal limits. In recent forms they have relief heights of about 0·7 metres (27 inches). However, pre-Paleozoic* stromatolites have a much greater amplitude than the recent ones, ranging from about 2·5 to 6 times as great. This points directly to the belief that tides in ancient times were generally much bigger than now. For this to occur the Moon must have been much closer to the Earth.

Rocks in the Swaziland System of south-east Africa show evidence of large tidal amplitude, and evidence provided by the oldest known metamorphic and granitic rocks by lead-isotope dating methods indicate that the Earth suffered some thermal event about 3·5 to 3·6 aeons ago. The effect of the Earth capturing the Moon at this time would give rise to sufficient tidal frictional heat to cause subcrustal melting. The melting would promote outgassing of the rocks and the gradual build-up of both an atmosphere and a hydrosphere. Any pre-existing terrestrial atmosphere and hydrosphere would have been lost before the new one was triggered off. Such a close encounter between the Earth and the Moon 3 aeons ago would also cause a temporary atmosphere to form on the lunar surface.

In the early years of the Earth, oxygen was locked away in chemical sediments, and free oxygen did not appear in the atmosphere until it later escaped via the oceans. If the hydrosphere and atmosphere began 3·5–3·6 aeons ago, life on Earth must have followed fairly quickly, for microfossils dated at about 3·2 aeons have been identified in ancient rocks.

Although the theories that account for the mechanism and methods by which the first free water that composed the oceans found its way into the basins must be speculative, no speculation is required to account for the presence of the water itself, for water is no rarity in cosmic space in any of its fundamental states as a solid,

* See Table A for the chronology of the geological time scale.

ERA	PERIOD	EPOCH	YEARS BEFORE PRESENT
CENOZOIC	QUATERNARY	RECENT	11,000
		PLEISTOCENE	5 TO 3 MILLION
	TERTIARY	PLIOCENE	13 ± 1 MILLION
		MIOCENE	25 ± 1 MILLION
		OLIGOCENE	36 ± 2 MILLION
		EOCENE	58 ± 2 MILLION
		PALEOCENE	63 ± 2 MILLION
MESOZOIC	CRETACEOUS		135 ± 5 MILLION
	JURASSIC		180 ± 5 MILLION
	TRIASSIC		230 ± 10 MILLION
PALEOZOIC	PERMIAN		280 ± 10 MILLION
	CARBONIFEROUS	PENNSYLVANIAN	310 ± 10 MILLION
		MISSISSIPPIAN	345 ± 10 MILLION
	DEVONIAN		405 ± 10 MILLION
	SILURIAN		425 ± 10 MILLION
	ORDOVICIAN		500 ± 10 MILLION
	CAMBRIAN		600 ± 50 MILLION
PROTEROZOIC	INFRA-CAMBRIAN		
	PRE-CAMBRIAN		
ARCHEOZOIC			4·6 AEONS*

*1 AEON 1000 MILLION YEARS

Table A The geological time scale

liquid or vapour. Meteorites similar in composition to planetesimal material contain water, and the variety known as carbonaceous chondrites contain water bound up in the form of the hydroxyl chemical groups. The rocks comprising the Earth hold vast quantities of water, and many of these rocks are composed largely of silicate material which contains hydrated crystals, so that water is incorporated as part of the natural atomic structure. Crystalline rocks contain from 4 to 8 per cent water by volume. This locked-up water can only be released by heat, and it is possible to account for all the waters of the hydrosphere in terms of the amount of thermal energy released by radioactive decay processes occurring inside the crust or by volcanism without invoking the energy brought about by the moon capture theory.

Thus the simplest picture is one where the Earth's hydrosphere was formed as a direct result of the 'sweating out' of water from the crustal rocks from energy derived by radioactive decay and/or volcanism. This began as a slow process which today is continued and added to by the additional action of the great subcrustal machine responsible for sea-floor spreading (see p. 63). The mid-ocean ridges are the sources of much fresh material which is drawn up from the depths to the surface where juvenile primordial water is then released.

The fact that the oceans are not saltier than they are is an indication that much of the free water now on the Earth's surface arrived at a comparatively late geological stage, and indeed many believe that the rocks forming the continental masses are at least ten times older than the water contained in the present-day ocean basins.

P.L.B.

2 DISTRIBUTION, COMPOSITION AND SOME PROPERTIES OF SEA-WATER

Distribution of Sea-Water

Most of the Earth's supply of free water is contained in the ocean basins and inland seas. The statistics of the world's oceans are impressive. They cover an area of 364 million square kilometres (140 million square miles) or about 71 per cent of the Earth's surface. The seas and oceans contain some 1,200 million cubic kilometres or 330 million cubic miles ($2{\cdot}75 \times 10^{20}$ gallons).

The world's present-day geological water contained in the hydrosphere bank is distributed by 97 per cent in the oceans, 1 per cent in the atmosphere and 2 per cent in ground water and the frozen ice-caps.* During the past ice ages the percentage distribution locked into ice-caps was much greater.

Composition of Sea-Water

Sea-water should not be imagined as just fresh water with mineral salts added; it is a biotic soup and resembles the soil in being fertile, enabling the phytoplankton (see p. 67) to propagate. It seems almost certain that the primeval oceans were the medium for the beginnings of all life on Earth.

The composition of sea-water, particularly its salt content, varies with temperature, latitude and the amount of fresh water brought down by rivers. Some of the inland seas, such as the Caspian,† are not formed of sea-water, and their chemistry is different. However, some of the other inland 'seas' have the highest salinity of all, such as the Great Salt Lake in the U.S.A. (salinity 280 parts per thousand). In the open seas the Arctic and Antarctic oceans have the lowest salt content in the upper regions owing to the continual influx of fresh water from the adjacent melting ice-sheets.

All the natural elements are represented in sea-water suspended in solution (see Table B). Sea-water contains about 35 grams per litre (5 ounces per gallon) dissolved salts of all kinds. Sea-water salts contain 89 per cent chlorides, 10 per cent sulphates and 0·2 per cent carbonates.

When discussing the salts of the sea, most people immediately think of sodium chloride ($NaCl$) and indeed this is one of the principal constituents, and the sea has been important as an economic source of 'salt' since earliest times (see also 'The Mineral and Chemical Harvest' pp. 183–91). The average salinity of sea-water is approximately 35 parts per thousand or, in more concrete terms, a

* It has been estimated that the amount of water in the Greenland and Antarctic ice-caps is about 25 million cubic kilometres (6 million cubic miles).

† The Caspian Sea has three times more calcium carbonate than the oceans, but is much less salty.

Table B Composition of sea-water at 35 parts per thousand salinity (in ascending order of the atomic weights of the elements)

Element	Symbol	Micrograms per Litre
Hydrogen	H	1·10 × 10^8
Helium	He	0·0072
Lithium	Li	170
Beryllium	Be	0·0006
Boron	B	4,450
Carbon	C	
(inorganic)		28,000
(dissolved organic)		2,000
Nitrogen		
(dissolved N_2)		15,500
(as NO_3^-, NO_2^-, NH_4^+)		670
Oxygen		
(dissolved O_2)		6,000
(as H_2O)		8·83 × 10^8
Fluorine	F	1,300
Neon	Ne	0·120
Sodium	Na	1·08 × 10^7
Magnesium	Mg	1·29 × 10^6
Aluminium	Al	1
Silicon	Si	2,900
Phosphorus	P	88
Sulphur	S	9·04 × 10^5
Chlorine	Cl	1·94 × 10^7
Argon	Ar	450
Potassium	K	3·92 × 10^5
Calcium	Ca	4·11 × 10^5
Scandium	Sc	<0·004
Titanium	Ti	1
Vanadium	V	1·9
Chromium	Cr	0·2
Manganese	Mn	1·9
Iron	Fe	3·4
Cobalt	Co	0·39
Nickel	Ni	6·6
Copper	Cu	23
Zinc	Zn	11
Gallium	Ga	0·03
Germanium	Ge	0·06
Arsenic	As	2·6
Selenium	Se	0·090
Bromine	Br	6·73 × 10^4
Krypton	Kr	0·21
Rubidium	Rb	120
Strontium	Sr	8,100
Yttrium	Y	0·003
Zirconium	Zr	0·026
Niobium	Nb	0·015
Molybdenum	Mo	10
Ruthenium	Ru	
Rhodium	Rh	
Palladium	Pd	
Silver	Ag	0·28
Cadmium	Cd	0·11
Indium	In	
Tin	Sn	0·81
Antimony	Sb	0·33
Tellurium	Te	
Iodine	I	64
Xenon	Xe	0·047
Cesium	Cs	0·30
Barium	Ba	21
Lanthanum	La	0·0029
Cerium	Ce	0·0013
Praesodymium	Pr	0·00064
Neodymium	Nd	0·0023
Samarium	Sm	0·00042
Europium	Eu	0·000114
Gadolinium	Gd	0·0006
Terbium	Tb	0·0009
Dysprosium	Dy	0·00073
Holmium	Ho	0·00022
Erbium	Er	0·00061
Thulium	Tm	0·00013
Ytterbium	Yb	0·00012
Lutetium	Lu	<0·008
Hafnium	Hf	<0·0025
Tantalum	Ta	<0·001
Tungsten	W	
Rhenium	Re	
Osmium	Os	
Iridium	Ir	
Platinum	Pt	
Gold	Au	0·011
Mercury	Hg	0·15
Thallium	Tl	
Lead	Pb	0·03
Bismuth	Bi	0·02
Radium	Ra	1 × 10^{-13}
Thorium	Th	0·0015
Protactinium	Pa	2 × 10^{-10}
Uranium	U	3·3

Table C Regional averages of silver, cobalt and nickel in the oceans expressed in micrograms per litre (after Schutz and Turekian)

Region	*Silver*	*Cobalt*	*Nickel*
Caribbean Sea	0·25	0·078	2·1
Gulf of Mexico	0·16	0·84	2·0
Labrador Sea	0·13	0·16	4·9
Northwest Atlantic	0·19	0·21	3·5
Northeast Atlantic	0·25	0·13	3·1
Southwest Atlantic	0·18	0·22	4·8
Southeast Atlantic	0·64	0·25	19·2
Indian Ocean	0·69	1·4	5·4
Central Pacific	0·34	0·75	20
East Pacific	0·23	0·18	5·5
Antarctic (East Pacific sector)	0·017	0·031	1·6

cubic kilometre of sea-water contains about 40 million tons of salt.* The oceans and seas contain enough salt to cover the continental land surfaces to a depth of 150 metres (500 feet).

Other important constituent compounds are magnesium chloride ($MgCl_2$), magnesium sulphate ($MgSO_4$), calcium sulphate ($CaSO_4$), calcium carbonate ($CaCO_3$) and potassium sulphate (K_2SO_4). Reference to Table B shows that many economic and precious metals are present in sea-water. Gold is present in quantities of approximately ·006 parts per million (38–40 pounds per million gallons of sea-water). The Caspian Sea alone receives 600 tons of gold via the river Volga each year. The problem of the extraction of gold from the sea has attracted the attention of men in modern times almost as much as did alchemy during the Renaissance. So far no process has been devised to extract the metal on an economic basis, but at least one modern national oceanographic expedition set out to investigate its possibilities (see p. 184).

Other elements such as magnesium and bromine have been commercially exploited, and at the present time 65 per cent of the world's production of magnesium metal is won directly from sea-water and over 5 per cent of magnesium compounds.

The ionic† composition of sea-water is an important factor in the

* Curiously the flesh of fish contains little salt.

† Hydrogen-ion concentration.

lives of marine animals. The pH value of sea-water is about 8, and even a change of 0·5 would likely bring about a disaster to much ocean life. It has been estimated that if only 1 per cent of the carbon at present locked away in the crustal rocks were liberated, a complete change would occur. If the pH value fell lower than 7, marine animals which form calcareous skeletons would stop growing, for calcium carbonate would not be precipitated.

The composition of sea-water does not appear to have changed much over the history of the oceans, and it is believed that some kind of chemical regulating mechanism is operating to maintain the ionic equilibrium; thus it follows that the carbon dioxide content of the atmosphere has been controlled within narrow limits. One likely candidate involved in such a control are the alumino-silicate clay minerals represented in the marine sediments which may serve as a buffer for tendency to change (see p. 66). One puzzling feature of the composition of sea-water is what happens to the dissolved silica in the form of silicic acid which pours into the oceans by rivers eroding continental silica rocks. Silica concentrations are always lower in sea-water than in river-water sources, therefore the sea must have a mechanism of extracting it rapidly from sea-water if the concentration becomes too high.

Silica is utilized by many marine organisms, chiefly diatoms, radiolaria and some sponges, for skeleton building. In some areas of the ocean it settles on the sea floor and is removed from use by marine organisms, but elsewhere control of the silica level appears to be biological in origin. It would appear that probably two mechanisms operate: a biological one for short term control, and one involving clay mineral sediments for long term (geological time-scale) control.

Carbonate of lime secreting organisms withdraw large quantities of calcium carbonate from sea-water to form shell and coral. Part of the withdrawn material accumulates in coral reefs, and part is taken up by the pelagic animals at moderate depths in the oceans. After the death of the organism, part of the calcium carbonate is redissolved.

Potassium salts are withdrawn from sea-water to combine with clay to form glauconite in the presence of organic matter. Potassium and iodine are selectively absorbed by certain species of seaweed, and until recent times seaweed was the chief economic source of iodine.

Sodium and chlorine appear to be the only elements that are

scarcely withdrawn at all except for small amounts deposited near the shores of land surfaces.

Some Properties of Fresh Water and Sea-Water

1 Unlike most liquids, water expands when it freezes. Thus ice floats rather than sinks.
2 Water has a high capacity* for heat storage (see 'The Ocean as a Global Thermostat', p. 34).
3 The specific heat of sea-water is less than that of fresh water.
4 The specific heat of water is higher than in any other liquid or solid. Steam and ice have only half the heat capacity of liquid water.
5 Sea-water is a better conductor of heat than fresh water.
6 Any sea-water on being cooled will always sink through water of equal salinity.
7 Fresh water reaches maximum density at 4° C. ($39\cdot2^\circ$ F.) and will always, at this temperature, sink through water at higher or lower temperatures.
8 Maximum density of average sea-water is reached at $-2\cdot2^\circ$ C. (28° F.) at its freezing point.
9 Temperature of sea-water varies between $-3\cdot3^\circ$ C. (26° F.) off Nova Scotia to 35° C. (95° F.) in the Persian Gulf and Red Sea.

P.L.B.

3 CONFIGURATIONS OF THE SEAS AND OCEANS

The greater part of the 71 per cent surface area taken up by the world's seas and oceans is accounted for by the four great ocean basins.

Pacific	165·5 sq. km. (millions)	63·9 sq. miles (millions)
Atlantic	82·1	31·7
Indian	73·6	28·4
Arctic	14·0	5·4
	335·2	129·4

* Capacity for heat is ability to store heat and is not to be confused with temperature.

The remaining marginal seas such as the North Sea, the Mediterranean, the Baltic, the Bering Sea, etc., comprise an area of 18 million square kilometres (10 million square miles). Apart from the above four general oceans, many recognize a fifth ocean, the Antarctic or Southern Ocean, which in area represents the southern parts of the Pacific, Altantic and Indian oceans approximately to a line below latitude 40° South.

Depths of the Oceans

The average depth of the oceans is 3,800 metres (12,500 feet). The deepest part of the ocean occurs in the Marianas Trench in the western Pacific and is 11,040 metres (36,200 feet) deep (Fig. 22). However, in thickness the oceans represent only one-sixteen-hundredth of the Earth's radius. If we depicted the Atlantic's greatest depth, 8,800 metres (29,000 feet), on a vertical scale of one centimetre, the width of the Atlantic must be represented by a measurement of no less than *c.* 6 metres (20 feet).

Structural Features of the Oceans

The oceans like the dry surface of the Earth possess well-defined structural features which bring about varying underwater landscapes and environments. The chief regions are designated the continental shelves, the continental slopes, the abyssal plains and the deeps, or trenches (Fig. 22).

The Continental Shelf

Many of the large continental masses have well-defined continental shelves* in their immediate off-shore waters. The shelves underlie only 7·5 per cent of the area occupied by the oceans, but more significantly they equal 18 per cent of the Earth's dry land area. The shelf area is defined by a gently sloping platform between the low spring tide to the 135–180-metre (445–595-foot) mark, but on a global basis the edge of the continental shelf ranges in depth from 20 to 550 metres (70 to 1,800 feet) and may be up to 1,500 kilometres (940 miles) wide as, for example, off the coast of Arctic Siberia. Along some of the continental masses the shelf is much narrower and in places almost entirely absent; one such place is along the shores of the eastern Pacific.

* The term continental borderline has also been used.

The channels cut by many great inland rivers can be readily traced along the underwater shelf, and there can be no doubt that the shelf areas once formed part of the continental dry land which is now submerged as a consequence of a depression in the continental mass and/or a rise in general sea level.* Although some of the outlying sedimentary material forming present-day continental shelves may be due to deposition during glacial periods, most of it belongs as part of the continental masses proper. The existence of a shelf along a continent with a fairly abrupt boundary at the junction with the continental slope indicates that the present level of water in the oceans is now greater than when the existing continental masses took shape.

From a geological aspect continental shelves comprise two types: those that have underlying sedimentary rocks, and those with underlying igneous or metamorphic rocks. The shelves may also be classified in other ways. Some are formed by geological uplift or upwelling of lava; some are the result of coral reef development. Others have been influenced by salt domes rising from below. Most, however, are due to simple deposition of sediment on top of igneous and metamorphic rocks, much of which has been removed from the adjacent continental mass by river action. The sediments comprising the deposits of 70 per cent of the world's shelf area have been laid down during the past 15,000 years since the last Pleistocene ice age† lowered the sea level.

Pollen samples from shelf seabeds show a succession of climatic changes beginning with tundra and then, about 12,000 years ago, spruce, which later gave way to oak and other temperate-climate deciduous trees, until the land was finally submerged by the rising sea. On the continental shelf of the North Sea, animal bones are regularly caught up in bottom trawls. Off the east coast of the U.S.A. remains of mammoth, mastodon, musk ox, horse, giant moose, tapir and giant sloths are regularly identified among bottom hauls.

From carbon-14 datings taken along the North American east coast it appears that the sea was near its present level *c.* 35,000 years

* During the ice ages the shelves were often bared to a depth of 150 metres (500 feet) or more below the present level of the oceans. The four principal ice ages of the Pleistocene can be traced out along present-day continental shelves.

† The last ice age is known as the Wisconsin in the U.S.A.; the New Drift in Great Britain; the Würm (W) in Central Europe; and the Weichsel in Northern Europe.

ago, and that it then began to recede *c.* 30,000 years ago. The level appears to have dropped by over 130 metres (430 feet) 15,000 years ago, but 6,000 years ago it rose rather rapidly to within about 5 metres (16 feet) of the present-day level.

Carbon-14-dated sequences from elsewhere in the world show a broadly similar pattern except where active zones of localized uplift have produced very marked differences. If the seas continue to rise as the ice-caps retreat, large areas of the Earth will eventually be engulfed.

The continental shelf area was the first part of the ocean bottom to be studied by man, and its first mention, as a feature, is attributed to Herodotus in about 450 B.C. Nowadays the shelves are extremely important for economic reasons, and present-day exploitation of the resources that lie under or above the bed of the seas and oceans is chiefly concerned with the shelf areas (see p. 177).

The Continental Slope

At the edge of the continental shelf platform, the seaward side makes a marked change in gradient, changing on average from 2° (or less than 2°) to about 5°. In some localities changes may be as abrupt as 15°, and in isolated instances almost cliff-like structures separate the slope from the lower abyssal plain.

The Abyssal Plain

Nearly two-thirds of the ocean floor lies at depths between 3,400 and 5,500 metres (11,200 to 18,150 feet) representing the gentle undulating feature known as the abyss, or abyssal plain. The name abyssal plain is somewhat of a misnomer, for in some areas the ocean bed profile is as rugged and scoured as a range of dry-land foothills or badlands. One of the most significant characteristics of the abyssal plain is the winding mid-oceanic ridges, which extend the length and breadth of the four great oceans (see pp. 53–4). The plain also contains minor ridge features and other analogous dry-land features such as escarpments, canyons and fracture zones, and a variety of submerged volcanic peaks known as seamounts. Some of these seamounts have remarkable flat table-tops which stop short of the surface by several hundred metres, and they are sometimes referred to as guyots (see p. 16).

The Oceanic Deeps or Trenches

It was a surprise to the early oceanographers to find that the deepest parts of the oceans were not found in the middle of the oceans as they expected. On the contrary, many of the deepest soundings have been located close to the continental shores or adjacent to chains of volcanic islands forming distinctive arcs.

The deeps are found in trenches or troughs in the seabed sometimes following the configuration of the adjacent island arcs so that the name *foredeep* is also sometimes applied to describe them. The separation of the deep from the general abyssal plain is about 7° or 8°, but the slope adjacent to the land side of the deep is usually considerably steeper. The deepest trench so far encountered is near the Marianas Islands* (11,040 metres; 36,200 feet) located in the western Pacific (see Fig. 22). Some of the Pacific deeps contain animal remains and minerals normally only deposited in shallower waters, and until recently these trenches were a mystery to marine geologists. It is now known that they play a significant role in the evolution of the ocean floor and in the 'continental-drift' theories of the wandering tectonic plate masses (see p. 55).

The Pacific Ocean

Including its bordering seas the Pacific Ocean occupies an area of about one-third of the total surface area of the world, representing an area one-eighth larger than the Earth's total land area. At its widest points, between the Bering Strait and the Antarctic continent, it stretches a distance of 15,000 kilometres (9,300 miles), and along the line of the Equator 16,000 kilometres (10,000 miles). Within its recognized confines the ocean basin contains 726 million cubic kilometres (174 million cubic miles) of water.

Most of the Pacific floor is abyssal plain whose average depth is 7,300 metres (24,100 feet) and exceeds that of the other oceans. One of the largest sea-floor features is the East Pacific Rise, also called the Albatross Plateau, which extends off the coasts of the Americas. In mid-Pacific lies the Hawaiian Swell 2,560 kilometres (1,600 miles) in length, on which rise the massive volcanic peaks of the Hawaiian Islands whose tallest peak towers more than 3,900 metres (13,000

* Discovered by the *Challenger* Expedition in 1872.

feet) above sea level, and whose total height in rising from the seabed is greater than Mount Everest's 8,708 metres (29,028 feet). Most of the islands which dot the Pacific have a volcanic base, and many are aligned in chains or arcs. During World War II, a U.S. naval officer, Harry H. Hess, discovered a new kind of volcanic cone while using depth-recording anti-submarine sonic devices. The ship's recording charts traced out flat truncated cones (Fig. 1) very much like the traditional pictures of a volcanic cone with the top neatly sliced off. During a period of two years while the U.S. *Cape Johnson* cruised in various Pacific war theatres, Captain Hess continued his wartime

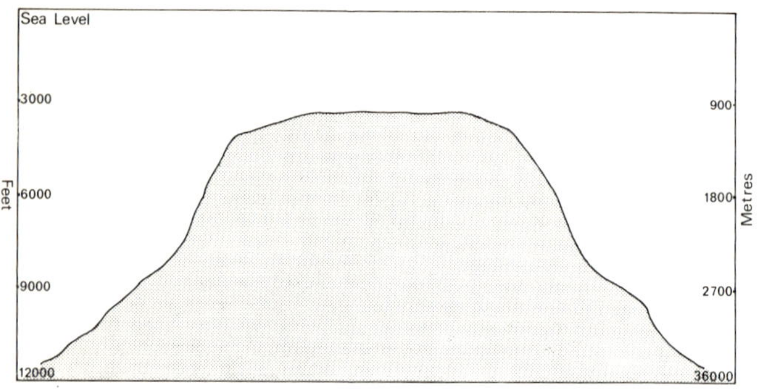

Fig. 1 The schematic profile of a typical guyot situated south of Eniwetok Atoll in the Pacific. The truncated platform is *c.* 15 kilometres (9 miles) across with a gently sloping rim to *c.* 1,260 metres (4,200 feet). The cliff-like edges flatten out to meet the ocean floor at *c.* 4,700 metres (15,600 feet).

oceanographic programme under the unsuspecting eye of his admiral. He charted the positions of twenty such flat-topped cones and later proposed that they should be named guyots after the famous Swiss oceanographer Andre Guyot. Nowadays many hundreds of guyots have been mapped in all the oceans, and a considerable number is to be found in the North Atlantic.

The flat tops of the guyots range in diameter from a few kilometres to more than thirty kilometres. They slope gently down to the abyssal plain at an angle of 20–25° (Fig. 1). Many lie under two thousand metres of water, but in earlier times their summits lay nearer to the surface.

The so-called seascarps are another characteristic feature of the Pacific floor. They extend in an east–west direction and are the result of faulting on a large scale; some have been traced out for a distance of 5,300 kilometres (3,300 miles), and they form impressive seabed relief scarps 300–1,500 metres (1,000–5,000 feet) high.

Perhaps the most striking features of the Pacific are the long trenches. These are sometimes called deeps, since they represent the deepest sections of the oceans. Some of the deeps are long and narrow, hence the usage of the description 'trenches', but it must not be imagined that they are similar in appearance to the shapes of the more familiar man-made trenches with vertical sides. In illustrations a false impression is often given owing to differences in horizontal and vertical scales shown in trench profiles (see Fig. 2). Most of the greatest deeps are associated with island arcs such as the Tonga–Kermadec Trench, 9,427 metres (31,109 feet); the Marianas Trench,

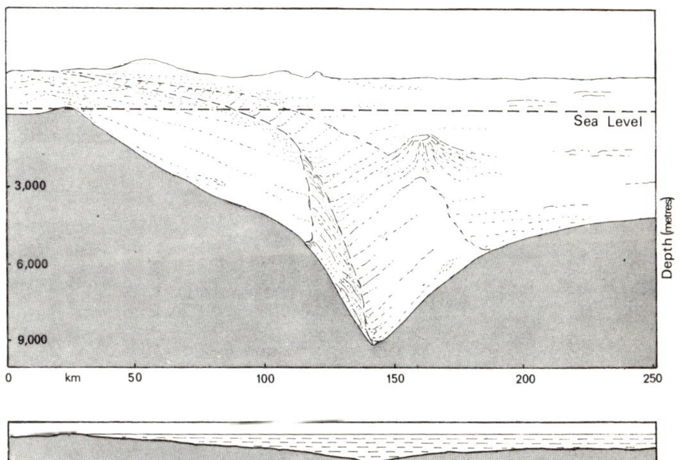

Fig. 2 The Tonga Trench in the Pacific Ocean. The upper figure, with the vertical scale exaggerated *c.* 10 times, shows the trench as viewed northwards from a position in the middle of the Tonga Islands. The lower figure shows the actual cross-sectional view without exaggeration of vertical scale. The guyot (or seamount) depicted on the eastern slope of the trench (upper figure) is one of the highest 'mountains' in the world in terms of its vertical rise from the sea floor to its submerged wave-worn summit (see p. 16 and Fig. 1).

11,040 metres (36,200 feet); the Japanese Kurile Trench, 1,055 metres (3,480 feet); the Sunda Trench, 9,427 metres (31,109 feet) and the Aleutian Trench 7,682 metres (25,350 feet) (Fig. 22).

Along the eastern side of the Pacific a great chain of trenches lies close against the mountainous margin of the Andes, and these have provided a significant evidence in unravelling some of the mysteries about the evolution of the ocean floor (see p. 55).

The Atlantic Ocean

The Atlantic covers about half the area of the Pacific, or rather less than one-sixth the total area of the Earth. Its widest parts are located in the temperate zones of either hemisphere. Towards the equator it is narrowed by the continental bulges of South America and Africa where their shapes suggest that the two continents fit together like complementary pieces in a jig-saw puzzle. This was the first clue that perhaps the separate continents once formed a single mass.

On the Atlantic seabed is hidden one of the most striking geological features of the Earth's crust. Indications of it were first found by the H.M.S. *Challenger* Expedition in 1873 when an anomalous rise was discovered in the mid-South Atlantic – this later became known as the Challenger Rise. Subsequent oceanographic expeditions found further extensions of it, and gradually the presence of a great longitudinal rise, or ridge, was traced out along the entire length of the ocean bed. Now this huge rise is known as the Mid-Atlantic Ridge (see Fig. 30). There are other ridges present in all the world's oceans. In British waters the Wyville-Thomson Ridge lies between Scotland and Iceland. All these ridges are interconnected on a world-wide basis and form a global network pattern.

The north-east Atlantic is also characterized by a remarkable mid-ocean submarine canyon, which begins between East Greenland and Labrador and extends southwards to be joined by a branch from West Greenland. After joining, the canyon extends farther southwards to latitude 37° North. Tectonic movements in the region of this submarine canyon caused the entire complex of seabed cables crossing it to rupture on 18 November 1929. These earth movements set into motion a colossal avalanche of sediments along the steep North American continental slope, which travelled downwards at a speed of 80 kilometres (50 miles) per hour until it finally rolled to a halt on the flatter abyssal plain below.

Unlike in the Pacific Ocean, lines of deeps, or trenches, are uncommon in the Atlantic, and the chief examples lie near the volcanic island arcs of the West Indies. The deepest Atlantic trench so far discovered lies near the island of Puerto Rico and is 8,800 metres (29,000 feet) deep. Another of the deeps – the Romanche Deep – extends across the Mid-Atlantic Ridge near its narrowest width.

Most of the marginal Atlantic seas lie towards the north-east. The North Sea and the Baltic are recent shallow seas whose origin is due to post-glacial submergence of the low-lying European continental land mass. The Mediterranean has several distinct basins, and it has a more complicated history; some of its deepest parts exceed 3,600 metres (11,900 feet).

The Indian Ocean

Unlike the Pacific and the Atlantic oceans the Indian Ocean is entirely enclosed by land to the north. Almost 60 per cent of its area forms an abyssal plain varying in depth between 3,600 and 5,500 metres (11,900 and 18,200 feet) and, apart from the Sunda Trench lying adjacent to Java, no other isolated deeps occur. As with the Atlantic and Pacific a number of well-defined submarine ridges network the basin. The principal one, known as the Carlsberg Ridge,* extends from the Red Sea and the Gulf of Aden to a point in mid-ocean where it divides into two separate ridges, one which continues into the southern ocean and connects with the Mid-Atlantic Ridge, and the other which crosses the southern ocean in an eastward direction between Australia and Antarctica and connects with the Pacific Ridge.

The Red Sea is one of the most interesting of the marginal seas of the Indian Ocean. The great African Rift Valley extends northwards along the median line of the Red Sea floor and marks the contact edges of the tectonic plates whose movements are now believed to have caused the continents to drift about (see p. 60). The Persian Gulf is an area which is slowly filling with sediments brought down by the Tigris and Euphrates rivers. Since Babylonian times great

* Named after the Danish brewery that has financed so much oceanographic research.

changes have occurred in the area, and the once-flourishing ports from which the ancients sailed to Egypt, India and the spice islands of South East Asia now lie many miles inland.

The Arctic Ocean

The Arctic Ocean literally sits 'on top of the world', and the greater proportion of it is perpetually shrouded in ice. In area it is about one-twelfth that of the Pacific, or some 14·2 million square kilometres (5·5 million square miles). At one time it was believed that the Arctic Ocean was a single deep water basin, but in 1948–9

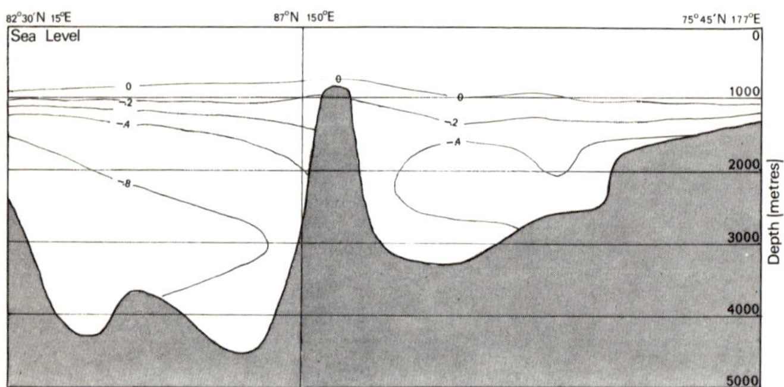

Fig. 3 Schematic profile through the Arctic Ocean showing how the Lomonosov Ridge divides the bottom Arctic water. Atlantic water is shown on the left, Pacific water (which enters the Arctic basin through the Bering Strait) is to the right. Figures show representative bottom-water temperatures in degrees centigrade.

Soviet polar expeditions discovered the Lomonosov Ridge* which extends like a giant submerged bridge from the Asian continental shelf to the continental shelf of North America near Ellesmere Land. This ridge is of great significance, for it plays a decisive role in the circulation and exchanges of water in the Arctic regions and diverts the bottom water into two distinct basins (see Fig. 3). This ridge is undoubtedly volcanic and is similar in origin to the Mid-Atlantic Ridge. It rises from the bottom to a height of 3,300 metres (11,000 feet), to a point within 960 metres (3,200 feet) of the surface.

* Named after the Russian naturalist Mikhail V. Lomonosov (1711–65).

Another lesser ridge also occurs in the northern part of the Greenland Sea. The greatest depth so far recorded in the Arctic Ocean is 5,360 metres (17,880 feet).

Another recent discovery in the shallower marginal Arctic seas are underwater mounds – with central cores of ice – called pingos. They were first noted in 1969 by a team of Canadian hydrographers on board the ice-breaker *John A. MacDonald*, and subsequently many others have been found over a wide area.*

It is believed that pingos represent offshore features such as unfrozen lakes which have been transformed by subsequent incursions of the sea. When the advancing sea reached them, the fresh lakewater mixed with brine so that the water forming the lake bottom cooled below the freezing point of fresh water and formed a core of fresh-water ice within the sediments of the former lake bottom.

Present ideas suggest that a lake 800 metres (2,600 feet) in diameter would develop into a pingo hill about 20 metres (66 feet) high. Pingos would grow slowly, and it is estimated that those in the Arctic regions have reached their present size over a period of about 5,000 years. This time requirement fits in very well with the evidence for a post-glacial advance of the sea, which occurred in the western Arctic about 5,000 years ago.

The Antarctic or Southern Ocean

Although the Antarctic, or Southern, Ocean is not officially recognized by geographers, since it really comprises the southern section of the Pacific, Atlantic and Indian oceans, many oceanographers nevertheless feel that the waters of these regions form a separate environmental zone with its own unique characteristics. Certainly from a marine biological point of view the Antarctic Ocean, or Southern Ocean (a title preferred by British and Soviet marine scientists), is one of the most important of all.

The waters of the Antarctic Ocean form a great unobstructed trans-circulation that is driven eastwards by the prevailing winds of the region. The boundary of the Antarctic oceanic region is ill-defined. One definition of its containment defines its northern limits

* Including traces of ancient pingos on dry land on what was previously part of the seabed.

at the Antarctic Convergence Zone which lies between 50° and 60° South latitude. The Antarctic Convergence is the name given to the area where the surface water flowing north converges with the surface water flowing south (see also p. 208). As a consequence of the meeting of the two kinds of water, there are marked changes in temperature, salinity and marine life (see 'The Environments', p. 64). In some quarters the Antarctic Ocean boundary has also been defined as the zone extending to the latitude 40° South which approximates to a line extending between the southern coasts of South Africa and Australia.

There is an extensive continental shelf area lying off the Antarctic continent, which is particularly extensive in the regions adjacent to Wilkes Land in the southern Indian Ocean sector, the Ross Ice-Shelf facing the southern Pacific Ocean and the Weddell Sea facing the Southern Atlantic Ocean. Most of the Antarctic continental shelf area is permanently covered by ice.

P.L.B.

4 MOVEMENT OF SEA-WATER

The movement of sea-water is what many people believe the science of oceanography to be principally about, and oceanographers are fond of making familiar analogies to demonstrate how important ocean circulation is. One famous Victorian pioneer oceanographer, when beginning a popular lecture to a lay public about oceanography, usually recited the following set piece:

> Oceanography has to do with the currents of water interchanged between tropical boilers – fired by the constant furnace of the Sun – and the polar refrigerators. Those rooms most exposed to the furnace are cooled down by iced water, while those more remote have their temperature raised by copious hot streams. Geology records many past contacts between the furnace and the ice-house in controlling the heating arrangements, and the many changes in the direction of flow in the hot and the cold water pipes . . .

It certainly is true that the movement of sea-water around the globe is one of the chief, seasonably variable, environmental factors that governs life on Earth. This movement is brought about by a number of independent factors but chiefly by the Earth's rotation, the radiant heat received from the Sun, and lunar/solar tides. The direction of movement of sea-water is not only influenced by the above factors but also by the general configuration of the continental land masses and to a lesser extent by the profile of the ocean bed.

When the Sun heats the surface water at the Equator, this warm water expands, with the result that the sea-water at the Equator is a few inches higher, creating a very shallow slope. As a result the surface water at the Equator streams 'downhill' towards the North and South Poles. The heavier, cooler water sinks below the warmer and tends to spread slowly along the bottom zone towards the Equator.

The interchange of warm equatorial water with cold polar water is a continuous operation which on a global scale never ceases. However, this picture of water movement is not as simple as it at first appears. Ocean currents are the result of the combination of many factors which are not always predictable, exact quantities.

In addition to heating the sea, solar radiation reaching the Earth also heats the atmosphere, which causes the movement of air masses, or simply wind. Winds originate owing to irregular heating of the Earth's surface by the Sun, causing convectional rising of air in the areas receiving more heat. These areas become regions of *low* air pressure, and cooler areas become regions of ***high*** air pressure where air descends and then moves into the low-pressure areas. Without exception, air always moves from regions of high pressure towards regions of low pressure. Air movements that move more or less parallel to the surface are called winds, while vertical up or down movements, such as those that exist at the centre of a convection system, are called currents. The circulations of all wind and ocean currents have been described as the parent and child of each other. All the initial thrust given to these movements of atmosphere and water is provided by the radiant energy of the Sun, three-quarters of which reaching the Earth falls on the oceans and seas. About a quarter of this energy is used in evaporating sea-water. This energy is in due course released as heat by condensation, bringing about atmospheric

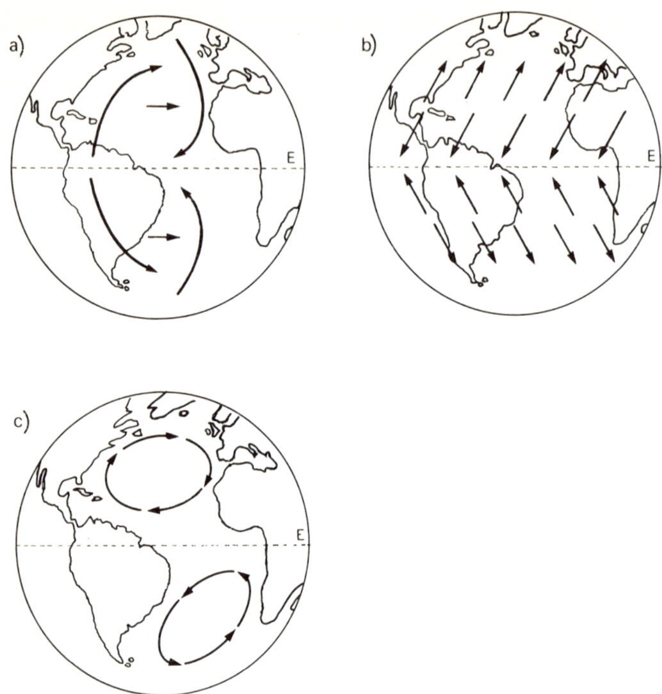

Fig. 4. a. The Coriolis Effect (or Force) is caused by the Earth's west-to-east rotation. This results in winds and ocean currents being deflected to the right in the Northern Hemisphere and to the left in the Southern Hemisphere. b. Wind belts are affected by the Earth's rotation and solar heat. In the tropics the trade winds cause movement of the surface waters westwards towards the Equator. In high latitudes movement is eastwards (the Westerlies). c. The combined results of three forces: the Sun's heat, the Coriolis Effect, and winds, bring about a clockwise circulation of ocean currents in the Northern Hemisphere and a counter-clockwise circulation in the Southern Hemisphere.

density changes which are the principal factor in producing wind. The winds push steadily over the oceans creating the major features of the current systems of the Earth.

The most consistent of the wind belts are the trade winds, so called because in the days of sail, ships made use of them in commercial enterprise. The 'trades' are situated at the fringes of the tropics and (owing to the Coriolis Force, see below) blow diagonally towards the Equator from an easterly direction in both hemispheres. If the Earth did not rotate on its axis, winds would probably move directly from high pressure to low pressure. However, they move at 60° to the perpendicular. This was shown mathematically by William Ferrel (1817–91). Ferrel's Law states that because of the Earth's rotation, Northern Hemisphere winds are deflected to their right, and Southern Hemisphere winds to their left. But winds blowing along the Equator are not deflected right or left.

Thus ocean circulation of sea-water is caused by the steady movement of winds in each hemisphere (the trades and also the westerlies). On both sides of the Equator the trade winds cause movement of the surface waters westward and towards the Equator. The Earth's rotation causes a deflection of these movements (like it does the wind) to the right in the Northern Hemisphere and to the left in the Southern Hemisphere.

The Coriolis Force

The force which causes the deflection due to rotation is nowadays named the Coriolis Force, or Effect, after the nineteenth-century French mathematician Gaspard Coriolis, who first analysed it in mathematical terms.* This force has far-reaching effects on all objects on the Earth's surface; but from an oceanographic point of view its chief influences are on the winds and ocean currents. Apart from one of its effects already stated, i.e. turning water to the right in the Northern Hemisphere and to the left in the Southern Hemisphere, it also causes water to tilt so that on the right side of a current the water is piled higher. This phenomenon has the result, for example, of influencing the Miami side of the Gulf Stream (see below) where the water is actually 60 centimetres (2 feet) lower than it is at Gun Cay on the Bahamas side, some 80 kilometres (50 miles) away. In the Sargasso Sea its waters stand appreciably higher than those on the circumference of the 'wheel' of the huge North Atlantic gyre, so that when ships leave the Sargasso Sea, they travel 'downhill'.

* According to one story Coriolis had the idea while playing a game of billiards, and shortly after he investigated the effects of compound centripetal acceleration.

Naval gunners have long recognized the effects of the Coriolis Force on the accuracy of aiming. A shell aimed at a target 32 kilometres (20 miles) distant travelling with a muzzle velocity of 750 metres (2,500 feet) per second would miss its target by over 60 metres (200 feet) unless a correction were applied. Ballistic missiles show the effect even more dramatically, and a missile with a flight time of about an hour across the North Pole would miss a target by 1,000 kilometres (625 miles) unless pre-corrections were given before launch.

Even on a smaller scale in the oceans and seas the effect of the Coriolis Force may have far-reaching consequences. For example, it has been demonstrated that the effect may have considerable influence on the mammoth 100,000-ton oil tankers at present plying the sea-lanes of the Earth. Some believe that perhaps more than one oil-tanker disaster – including the *Torrey Canyon* episode which seriously polluted the beaches in southern Britain – has been the direct result of the inability to correct the horizontal force perpendicular to the direction of the ship's velocity.

In a lighter vein, the Coriolis Effect is often cited as causing bath water in the Northern Hemisphere to run out clockwise while in the Southern Hemisphere counter-clockwise, but detailed investigation has exploded this long-cherished myth. Investigations found that other factors such as those involved in the construction of the bath are far more influential in determining the direction taken by such a small volume of water.

P.L.B.

OCEAN CIRCULATION

Gyres and Currents

Oceanographers term the great whirlpool patterns of water circulation, gyres* (Greek: *gyros*, a circle). There are six major gyres operating, one in either hemisphere of the Pacific, Atlantic and Indian oceans. Separate parts of the gyres are called *currents, drifts* or *streams*. Currents which flow away from the Equator are warm currents, those that flow towards the Equator are cold. Currents

* Or gyrots.

which move in the latitudes of the prevailing westerlies become broader and slower and are generally known as West Wind Drifts.

Pacific Ocean Circulation

The North Equatorial Current of the Pacific is the largest westerly running current on Earth with no barrier to deflect it in its 14,500-kilometre (9,000-mile) sweep from the Panama Isthmus to the Philippines. When it finally reaches the barrier of the West Pacific Islands, it swings northwards to become the Japan Current, better known as the Kuroshio or 'Black Current', because of the deep indigo blue of its water. The Kuroshio is the Pacific's counterpart to the Gulf Stream, bringing warm water to the coastline of Japan, but because the current is more slow-moving, it does not have quite the same effect as the Gulf Stream. East of Japan, the Kuroshio becomes the North Pacific Current and is forced eastwards by the icy-cold Oyashia Current flowing from the Bering Strait, infusing into the North Pacific chill Arctic water. By the time the North Pacific Current reaches the American West Coast as the California Current, it has cooled considerably and provides the west coast of North America with a temperate maritime climate. Farther south the California Current completes the gyre circulation by joining up with the North Equatorial Current.

In the South Pacific, the South Equatorial Current is a more diffuse and broken flow of water than its counterpart in the north, owing to the numerous islands it encounters in its push across the ocean.

By far the most important current of the South Pacific is the Humboldt, or Peru, Current. It originates in the cool waters of the Southern Ocean, and because the prevailing winds of the eastern coastline of South America blow offshore, surface water is replaced by deep water loaded with nutrient mineral salts which, after reaching the Peruvian part of the current, bring about super-abundant marine life. Here phytoplankton, zooplankton, fish and seabirds (see also p. 209) breed in concentrations unsurpassed elsewhere in the world. Among the seabirds the cormorant population is counted in millions, and they are maintained by the vast schools of anchova fishes. In turn the seabirds' droppings give rise to the rich Peruvian guano deposits. In terms of value to man, the economic (food) benefits of the Humboldt Current closely rival the thermal (social) benefits brought by the Gulf Stream to Western Europe.

Atlantic Ocean Circulation

The South Equatorial Current of the Atlantic flows westwards until it is split into two branches by the projecting coast of Brazil. The southern branch follows Brazil southwards as the Brazil Current, then it moves across the ocean as a West Wind Drift and returns northwards along the African west coast as the Benguela Current. The Benguela Current is the Atlantic counterpart of the Humboldt Current, and again its nature is brought about by offshore winds and the upwelling of nutrients. Although its life is not quite as prolific as that of the Humboldt Current (Peru Current), it nevertheless supports a population of 20 million seabirds – cormorants, gannets, penguins and gulls – which alone consume an estimated 6–7 million tons of fish annually. The northern branch of the Benguela Current finally crosses the Equator and completes the gyre with the North Equatorial Current.

The second branch of the North Equatorial Current flows westwards until it reaches the West Indies where it again splits into two branches. One branch flows northwards, while the other – the Caribbean Current – pushes into the Gulf of Mexico from where, warmed and enlarged, it emerges through the Strait of Florida as the Gulf Stream (see 'Nature of the Gulf Stream' below). It then flows along the coastline of the United States past Cape Hatteras where it joins the northern branch which divided off near the West Indies. Together they bend eastwards at about 40° North latitude and move towards Europe as the West Wind Drift. Important branches fan out along the shores of Iceland, the British Isles and Scandinavia, providing these regions with climates which are noticeably warmer than the average for this latitude. Another branch turns southwards as the Canary Current to complete the gyre.

Complete ocean circulation is also sometimes called an *eddy*. Inside the slow whirlpool of the North Atlantic Eddy, in the calm air of the Horse latitudes,* great masses of floating seaweed accumulate to form the Sargasso Sea.

* The name given to regions of calm and light variable winds. The origin of the name is supposed to hark back to the days of sail when vessels on the America run sometimes disposed of their cargoes of horses if becalmed for long periods.

The Labrador Current is a cold current that comes down from the Arctic into the North Atlantic. Flowing from Baffin Island it brings with it icebergs, pack-ice and icy water which chill the coasts of Labrador, Newfoundland and the coastline of New England as far south as Cape Cod. Beyond this point its cold, dense water sinks below the surface. The fogs off Newfoundland are a direct result of the cold water from the Labrador Current meeting the warm, moist air blown over it from the nearby Gulf Stream. Many vessels have collided with icebergs owing to the poor visibility caused by fogs.

Indian Ocean Circulation

The pattern of currents circulating in the Indian Ocean changes according to the seasons. From August to October the South-West Monsoon drives an eastward-flowing current, but during the winter months the monsoon reverses its direction and sets up a current flowing westwards.

In the Southern Indian Ocean, the West Australian Current is a cold current flowing along the coast northwards and originates in Antarctic waters. The Agulhas Current is a temperate current flowing south-westwards along the South African coastline.

Arctic Ocean Circulation

Since much of the Arctic Ocean is permanently covered in solid ice or by broken ice-drifts, Arctic surface currents may be traced directly by the drift of ice-floes (Fig. 25). There appear to be three major circulation systems. One system originates in the Bering Strait and moves across the North Pole. One of its minor branches reaches Ellesmere Island, then the main current flows towards the Greenland Sea and into the North Atlantic via the East Greenland Current. A second circulation is centred in the Beaufort Sea, forming a zone of stagnation (like an Arctic Sargasso Sea) in which the ice-floes slowly gyrate in a continuous closed, clockwise circulation. The third circulation divides into a number of diversionary circulations adjacent to the Siberian coastline; one branch moves towards the Greenland Sea and joins the current moving from the direction of the Bering Strait.

Antarctic Ocean Circulation

The flow of the Antarctic Ocean (Fig. 26) is the only complete uninterrupted west to east movement of water on Earth. However, some of its waters are diverted towards the north, giving rise to the West Australian Current and a current flowing into the Tasman Sea between Australia and New Zealand. It also contributes some water to the Agulhas Current along the South African coastline. At its narrowest constriction in Drake Passage, between South America and Antarctica, its estimated flow equals some 150 million cubic metres (5,290 million cubic feet) per second; between South Africa and Antarctica its flow is 190 million cubic metres (6,700 million cubic feet) per second; and between Tasmania and Antarctica 180 million cubic metres (6,350 million cubic feet) per second. The differences in volume are accounted for by its mixing and diversion into the southern parts of the three great oceans which border it.

Nature of the Gulf Stream

One of the first scientific studies of ocean circulation was made by Benjamin Franklin about 1769. Franklin had noted that some American ships took two weeks less than British ships to cross the Atlantic, and discussing this with a cousin – a Nantucket whaling captain – he was told that American captains had discovered that by steering a certain course, a current could be encountered which added about three knots to the speed of any vessel travelling to Europe. On their way back, the east-moving current could be avoided by sailing a much more southerly course. Nevertheless, the Gulf Stream was no new discovery made by the American captains, for it was known from the time of the early Spanish and Italian sailors, and the British seamen of Hakluyt's period knew it well.*

Franklin became so intrigued that he published a chart of the current and called it 'Gulf Stream'. However, the name seems to have been coined at a much earlier date. Since the Gulf Stream forms one of the most important currents of ocean circulation, it has been investigated in great detail.

Born of the North Equatorial Current (see above) via the Gulf of

* *The Principal Navigation Voyages, Traffics and Discoveries of the English Nation*, by Richard Hakluyt, published 1598–1600. (Modern edition, 'Everyman Library', Dent, London).

Mexico, the piled-up water flows past Florida with a surface temperature of about 27° C. (80° F.) but lower down with a temperature of about 10° C. (50° F.). Its volume at this point is over twenty-five times greater than all the rivers of the world. It then flows northwards – deflected towards the North American coast by the Bahamas – forming a huge stream of water 65 kilometres (40 miles) wide and 600 metres (2,000 feet) deep. However, its flow is complicated by eddies and counter-eddies whose pattern is continually changing. When it reaches the Grand Banks off Newfoundland, the boundary between the cold Arctic water and the water of the south is so abrupt that a difference of 10° C. or more may exist between temperatures taken at the bow and the stern of a vessel crossing it.

In 1969, a new 130-ton submersible craft appropriately named *Ben Franklin* (Fig. 27), designed by the famous Swiss engineer Jacques Piccard and built by the Grumman Aircraft Corporation, made a unique 2,400-kilometre (1,500-mile) submerged drift to explore its nature. During the course of this drift, the *Ben Franklin* varied its depth between 90 metres and 600 metres (300 feet and 2,000 feet), while following the stream from the Strait of Florida to a position east of Cape Cod, Massachusetts. During this slow drift northwards at 3½ knots, thousands of photographs were taken of the sea floor, and large areas were delineated for the first time. One surprising result of the voyage was the non-detection of the so-called deep scattering layers believed to be caused by nocturnal migrations of marine animals (see p. 69).

Deep Sea Currents

Whereas surface currents owe their motion principally to energy supplied by the wind, deep ocean currents are set in motion when surface waters become cooler and denser than the water below and thus begin to sink. Sea-water changes density in two ways, by an increase in dissolved salts and through a decrease in temperature. When salt water freezes, the salt is literally squeezed out of the water, and any unfrozen water remaining near by becomes considerably saltier. In the oceans, near the poles, two deep-sea currents are triggered by the increased saltiness of unfrozen water, which becomes the densest, coldest water of the entire oceans and slides down off the continental shelves into the deepest pockets of the

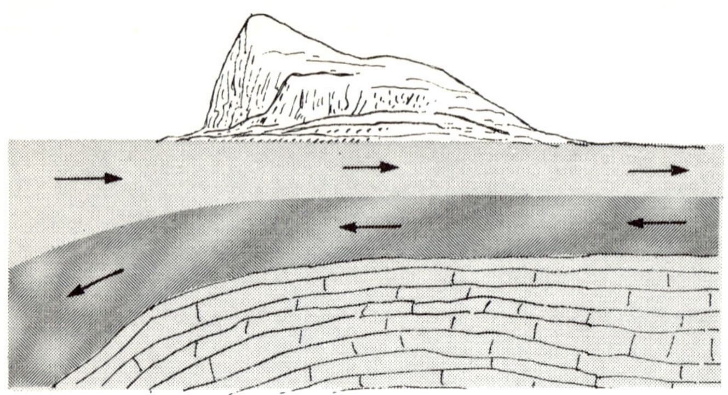

Fig. 5 The inflow and outflow of sea-water in the Mediterranean/Atlantic through the Strait of Gibraltar. Because of rapid evaporation in the Mediterranean, water becomes more salty than in the Atlantic and therefore sinks and flows outwards to be replaced by less salty water.

ocean floor. Although much of the equatorial Atlantic deep water has originated from the north, another important deep layer is due to the salt-laden inflow of the Mediterranean (see Fig. 5).

The global pattern of deep water currents is gradually being mapped. An equatorial undercurrent was found in the early 1960s in the Indian Ocean, moving at 2 to 3 knots, and its presence was first noticed by long-line tuna fishermen. It consists of a narrow eastward-moving stream with a core 50 to 100 metres (160 to 300 feet) below the surface. In the Atlantic a similar current was confirmed by the Soviet research vessel R. V. *Mikhail Lomonosov* at 50 metres (160 feet) depth, travelling at 2·25 knots, but first indications of this were found by the cable layer *Buccaneer* back in 1886. In 1957, International Geophysical Year scientists discovered the existence of a counter-current 2,700 metres (9,000 feet) beneath the surface of the Gulf Stream, travelling in the opposite direction at about 13 kilometres (8 miles) per day.

Much of the water in the deep sea has remained there for some time. By using various modern dating techniques, it is possible to estimate the ages of 'parcels' of trapped water. One such method is by the determination of its oxygen content. When water starts descending at the surface, it has a relatively rich supply of oxygen; as it sinks, this is gradually used up by the animals living in it and by the process of oxidation of dead organic material. The rate at which sea-water is depleted of oxygen is known from experience. By measuring a given sample of water in which the oxygen is depleted, an estimate may be made of its age, or rather of the elapsed time since it left the surface. World-wide water sampling using this method of ageing has shown that the 'youngest' sea-water occurs in two regions: round the Antarctic continent, and in the western North Atlantic. All the oceans of the world eventually receive their deep water from these two 'gestation' regions. Currents carry it first into the Indian Ocean and then in due course into the Pacific. The oldest water carrying the least amount of oxygen occurs in the abyss off the coast of Peru (Fig. 22).

Sea-water may also be aged by using the carbon-14 dating technique.* Although it is now known that carbon dating is subject to some uncertainty, particularly in long-term dating, it provides a more accurate dating method than the depletion-of-oxygen method. Using the carbon-14 technique, some of the water in the deeps of the South Pacific shows ages ranging from 1,350 to 1,910 years – implying that the deep-sea currents which have brought the water to these areas move at a velocity of about 15 kilometres ($7\frac{1}{2}$ miles) per year. Both these measuring techniques are used in the shorter term to obtain estimates of primary plant production (see p. 206).

Atlantic water deeper than 2,000 metres (6,500 feet) shows a range of ages between 50 and 1,000 years. Deep water in the Mediterranean and other near-by enclosed basins takes much longer to replenish than that in the open seas. The deep waters at the bottom of the Black Sea remain in isolation for very long periods, perhaps anything between 2,500 and 5,000 years.

The understanding of the pattern and movement of deep-sea currents is of particular importance today owing to global pollution (see p. 235). Since man began the industrial revolution,† the seas have been a natural dumping ground for toxic wastes. Many of these

* In this method use is made of the differences in carbon isotopes. The two main forms of carbon are carbon-14 (atomic weight 14) and carbon-12 (atomic weight 12). Carbon-14 decays to carbon-12 at a certain rate so that after a given interval an exact ratio of the two is present as a mixture. In any substance containing carbon (such as sea-water) the ratio of the amount of two isotopes can be measured very accurately.

† To a lesser extent, of course, since the time man evolved as a dominant species!

toxic wastes can be safetly diluted by sea-water and/or reduced to other less toxic materials through natural bacteriological processes. However, this cannot be done with radioactive substances. Many of these substances decay at much slower rates than the circulation rates of ocean currents so that still active material may sometimes be brought back to the surface with consequent great risk to life. An intensive study of ocean circulation is necessary so that best use can be made of those areas of the seabed where circulation is minimal, and where perhaps dangerous solids may be permanently disposed of. Such likely places are the deep ocean trenches (see p. 15).

The Ocean as a Global Thermostat

For the globe as a whole, the ocean is the great stabilizer of temperatures and is able within limits to maintain an equilibrium of life on the Earth's surface.

Sea-water is an excellent absorber and radiator of heat. Owing to its enormous heat capacity, it can continue to absorb great quantities of solar radiation without becoming unbearably hot, and conversely it can lose much of its heat without seemingly becoming too cold. Because of these qualities, sea-water acts as a gigantic heat storage battery for the Earth.*

Through the agency of ocean currents and the atmosphere, heat and cold may be distributed over thousands of miles. The ocean plays the most important role in this since, unlike the atmosphere – which is susceptible to sudden or excessive fluctuations of temperature – the oceans change temperature at a comparatively slow rate and therefore dominate the air.

In the traditional style of using familiar analogies, the British meteorologist C. E. P. Brooks once described the North Atlantic as a great bath with one hot tap and two cold taps. The hot tap is represented by the Gulf Stream and the cold taps by the East Greenland Current and the Labrador Current.† The adjustment of the three taps determines the surface temperatures in the eastern

* Temperatures in the English Channel vary from 8° C. (47° F.) in winter to 16° C. (61° F.) in summer. At the Florida Keys (Gulf Stream) they vary from 21° C. (70° F.) in winter to 28° C. (82° F.) in summer.

† The cold Labrador Current makes the water so cold that along the coast of New England, bathing is extremely uncomfortable even during the summer.

Atlantic. If only a minor change of warming occurs, say only a degree or so, this has far-reaching economic consequences on the amount of snow-cover in Europe. Less snow means earlier ploughing and a better chance of a good harvest. Meteorologists have learnt that long-range weather forecasting needs to reckon with the prevailing pattern of adjacent ocean temperatures.

It is not only the climate of Europe that is markedly influenced by ocean currents. In South America the cold Humboldt Current creates an arid desert-like climate along the eastern shore. The onshore breezes that carry towards the hot land in the afternoon are formed of cool air that has lain over a cool sea. The proximity of the high Andes Mountains causes this air to cool even more, so that there is little condensation of water vapour. On occasions when the normally chill Humboldt Current is deflected away, a warm current of tropical water called *El Nino* replaces it, and the downpours of rain which result sweep the dusty Peruvian hillsides for months, bringing disaster to homes and crops. A similar phenomenon occurs along the coast of South-West Africa normally influenced by the cool Benguela Current. In the Indian Ocean the monsoon gathers summer rain from the sea for India, and when this fails, it spells economic disaster for the entire subcontinent.

One of the best examples of the overall influence of the oceans on climate is in the contrast between the Arctic and Antarctic regions. In Antarctica the land mass is surrounded entirely by an ocean and bathed by seas of uniform coldness which prevent any warming influence reaching its interior lands. Only the hardiest life can survive this climatic rigour in spite of long hours of summer sunshine, where occasionally the radiant heat from the Sun allows polar travellers to strip to the waist to cool off. In contrast the Arctic has no immediate land mass at the pole, and the influence of the warm Atlantic penetrates right into the Arctic region, creating an environment a few degrees warmer, which enables the Arctic to support a wide variety of plant and animal life at extremely high latitudes – the equivalent of which in the south is barren ice-desert.

Ice Ages and Ocean Circulation

It seems certain that climatic changes which have occurred during the history of the Earth have been directly influenced by variations in the temperature of the ocean and its gyre circulations. This is disregarding climatic changes brought about by the redistribution of the continental masses occurring through continental drift plate tectonics (see p. 63).

The mechanism behind the recurrent ice ages has long been a puzzle to science. Numerous theories have been proposed to account for variations in the Earth's temperature. Theories involving a terrestrial origin range from the precessional wobble of the poles to variations of dust and carbon dioxide in the Earth's atmosphere. Volcanic eruptions are capable of releasing great quantities of dust into the atmosphere, which may remain there for years before slowly falling back to the ground. Volcanic dust particles have been found at altitudes of 50 kilometres (30 miles), and some scientists maintain that during periods of extensive volcanic activity, sufficient dust may be released to blanket the Earth and prevent solar radiation from reaching its surface. Carbon dioxide released by volcanoes might also give rise to a 'greenhouse effect'.* The balance of carbon dioxide in the atmosphere is important in relation to plant life in the sea which is the starting point in the great food chain pyramid (see p. 205).

Astronomical theories range from the Earth's encounter with comets to the cyclic variation in the Sun's output of heat radiation. The Sun certainly has a regular sunspot cycle of about eleven years, but this brings about only a minute variation in temperature (if any at all). Although theories have been put forward suggesting that the Sun varies over very long periods – for example the same as the ice-age cycle during the last million years, there is much stronger support for the idea that the Sun has not changed over the past 5,000 million years – longer than the estimated age of the Earth (4·6 thousand million years). Ice ages can be traced back into pre-Cambrian times, which indicates they are not a new phenomenon involved with recent continental drifting. No theory yet put forward is entirely satisfactory, but whatever the cause of the initial triggering mechanism, we know from an oceanic point of view that only a

* Carbon dioxide in the air allows the Sun's short light waves to penetrate to the Earth's surface. As the ground is heated by radiation, it sends back long-wave (heat-wave) radiation into the atmosphere, but the long waves cannot penetrate the carbon dioxide layer, with the result that a heat trap like that in a glasshouse or greenhouse is formed.

global fall of 2·2° C. (4° F.) in ocean temperature is necessary to bring about extensive glaciation. Thus it is seen that the temperature of the oceans holds life on Earth in a delicate thermal balance.

Some Atlantic seabed cores show that between 10,000 and 15,000 years ago the Atlantic experienced a sudden warming. The evidence is provided by layers of marine organisms which although formerly abundant in the Atlantic were suddenly replaced and covered by a layer of other varieties known to require higher temperatures. What is significant is that there was no transition between the two varieties; the division was sudden and clear-cut. Two American oceanographers, M. Ewing and W. L. Donn, decided to look into the cause of this sudden warming. One of the most puzzling features was that the abrupt change came at a period when the climate was very cold, shortly after the advance of the glaciers had stopped.

After examining various possibilities, Ewing and Donn finally concluded that a possible cause might be explained by supposing that the surface of the Arctic had once been open. If this happened, it would allow great quantities of cold water to pour down into the Atlantic and cause it to cool. If later the Arctic then froze over, the influx of cool water would suddenly diminish and allow the Atlantic to become warmer.

It is clear that an open Arctic Ocean, free of ice, would have significant consequences on the climate of the Northern Hemisphere if not over the entire globe. The alternate cycle of an open and then a closed Arctic Ocean would act like a thermostatic valve to the Atlantic and its neighbouring continents.

To follow up the consequences of their ideas, Ewing and Donn prepared weather charts to study the effect on the atmosphere. It soon became apparent that an open Arctic, influenced by warm Atlantic currents, would unload a tremendous volume of water vapour into the polar atmosphere. This would give rise to blizzards and periods of prolonged snowfall, finally leading to glacial conditions which would spread across and blanket North America, Europe and Siberian Asia.

But the build-up of ice on the continental masses would gradually rob the oceans of their water resources, bringing about a drop in sea-level. This, in turn, would slow the flow of warm water into the Arctic region, and when the sea-level reached the shallow 1,600 kilometre (1,000 mile) long bank which separates Greenland and Norway, glaciation would halt, for no longer would the warm Atlantic water reach the Arctic. Soon after a thaw would begin.

Nevertheless, the Ewing–Donn theory does not explain the initial mechanism which triggered the recent Ice Age, nor does it account satisfactorily for all the associated phenomena observed which are consequences of it.

It is, of course, not correct to apostrophize or simplify the events which occurred in the Pleistocene and think of it simply as a recent Ice Age consisting of a few separate periods of glaciation interrupted by warmer inter-glacial periods. From evidence of deep-sea cores it seems that at least some fifteen major phases occurred with the probability of many lesser ones. During the Pleistocene the temperatures rose above those prevailing in recent times, and the climatic changes brought about on both sea and land were far-reaching. On dry land the geological record of the Lower Pleistocene is woefully incomplete. In tropical and equatorial regions remote from glaciation, except at higher altitudes, there are many indications of periods of increased rainfall during the Pleistocene which are referred to as pluvials. Much discussion has centred on how the so-called pluvial periods might be synchronized with the chronology of the glacial sequences. Evidence has been put forward that in some areas there were two wet cycles and two dry cycles for each glacial/interglacial oscillation. This repeating pattern could have had far-reaching effects on prevailing weather patterns, on landforms and on human and other biological evolution.*

The results of the core drilling programme (Legs 28 and 29) by the *Glomar Challenger* in Antarctic waters during the early 1970s are likely to change many of our ideas about what has occurred in the last 20 million years or so of geological history. It was once believed that the Antarctic Ice Cap could not be more than 5 to 7 million years old, but cores now provide positive evidence that some kind of ice cap has been in existence for at least 20 million years. Five million years ago the ice cap had reached its maximum extent some 320 to 480 kilometres (200 to 300 miles) farther north than it now lies. Rapid melting reduced it to its present size, and it has remained

* Quaternary geology and the consequences of the recent Ice Age are covered in more detail in the companion volume *The Earth in Colour* (in preparation).

more or less stable ever since. The rapid melting, however, must have caused a rise in sea level of some several tens of feet. This would have affected the pattern of the Antarctic continental shelf which controls the very cold Antarctic bottom waters. In turn, these bottom waters regulate the circulation and the sedimentation pattern in the depths of all the world's oceans. Thus the retreat of the Antarctic ice must have had wide-reaching global repercussions. One consequence of this new knowledge is that it perhaps explains why sediments between 10 and 40 million years old have not been found in many of the deep ocean basins. Although the recent ice ages may certainly be accounted for by the Ewing–Donn theory, what about those in the past long before the Atlantic Ocean was formed? Wandering continents could account for them by shifting the flow of warm equatorial currents. Even today the Atlantic owes its favourable warm currents to the fact that the gap between North and South America is closed by the narrow Isthmus of Panama. If there were a channel here, the South Equatorial Current would drive straight through and escape into the Pacific, resulting in no warming Gulf Stream for Europe.

During the nineteenth century when the idea for the Panama Canal was first mooted, it was criticized on the grounds that it would change the heat distribution of the ocean waters by allowing the South Equatorial Current to escape into the Pacific. However, the canal is constructed as a series of different water levels linked by locks, but even so it is doubtful whether a through canal such as the Suez would be large enough to upset the flow of the North Atlantic circulation.

More recent ideas for damming the Bering Strait and the Strait of Gibraltar have also been criticized for the wide-sweeping climatic changes they might bring about on huge areas of the Earth. If the Bering Strait were dammed off, one effect would be the heating up of the Arctic which would allow year-round navigation of the normally ice-bound coast of Siberia and Alaska.

In historical times it is probable that the flow of the Gulf Stream has undergone some changes. If we look back 1,000 years, we find traces of Viking settlements in Greenland and Newfoundland when the North Atlantic climate was much more favourable than in the period of cold from 1400 to 1900 that followed it. From 1900 we again appear to be entering a warm phase. The world's glaciers are rapidly on the retreat, and if they continue to do so indefinitely – unleashing billions of gallons of melt-water into the oceans – many of the world's most important seaports including London, New York and Tokyo will one day sink below the waves. But before this happens, it is likely that the thermostatic valve will again be tripped by Nature, and the process reversed.*

Tides

The tides of the oceans are movements of water brought about by the gravitational pull of the Moon and to a lesser extent by the gravitational pull of the Sun. The combined pull of the Moon and the Sun on ocean water causes a wave-bulge of water to occur on both the nearest and farthest side of the Earth opposite the direction of pull (Fig. 6).

Because of the Earth's rotation (in 24 hours) typically two high and two low tides occur within 24 hours 51 minutes, since this is the period of the Earth's rotation in relation to the Moon's rotation round the Earth (29·5 days).

If tidal waves of the ocean did not experience friction against the irregular-shaped ocean bottoms and the resistance of the continents, there would be a tide every 12 hours 25 minutes due to the Moon, and one every 12 hours due to the Sun. However, owing to a great variety of interferences and obstacles, the globe as a whole experiences a wide range of tidal phenomena.

The most common are semi-diurnal tides. They change direction four times a day, rising from zero to a maximum and returning to zero twice during the 24-hour period. The Atlantic Ocean shoreline has very characteristic semi-diurnal tides which shift 51 minutes in the start of the phases each day owing to the Moon's 29·5 days' revolution round the Earth (as explained above).

Some parts of the world experience only one high and one low tide each day. Some coasts have tides of over 12 metres (40 feet), like the Bay of Fundy (Newfoundland), while others like the Mediterranean have a tidal range of only a few centimetres. What

* Whether indeed this has already occurred during the past decade or so is subject to conjecture. At the present time there is insufficient evidence available to confirm the onset of a long-term colder trend, and many glaciers appear to be still in retreat.

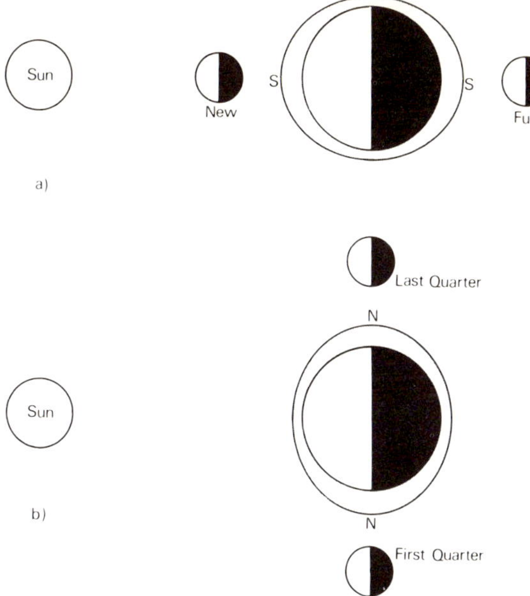

Fig. 6 The tidal bulge and spring and neap tides:
a. Spring tides (S) – greatest amplitude – occur when the Sun and Moon are lined up at the time of New and Full Moon.
b. Neap tides (N) – lowest amplitude – occur when the Moon is at First or Last Quarter.

determines the frequencies of tides is the shape of the tidal water basin, the shape of the shoreline and the movement of winds.

Twice each month during the periods of the New and Full Moon, when the Sun, Moon and Earth are in syzgy,* the tidal amplitude is greatest, and *spring* tides occur. Twice each month, when the Moon is at First or Last quarters, the pull of the Moon and the Sun oppose each other, and therefore the tidal amplitude is least, and *neap* tides occur (Fig. 6).

Unusual tidal effects occur where two currents run into each other, and the result may be a tumultuous 'tide-rip'. In narrow

* Syzgy = three bodies in a line.

passages between islands, or between an island and the mainland, tides may flow or ebb with considerable force at speeds up to 15 knots. Tide-rips also give rise to whirlpools as in the Strait of Messina (between Italy and Sicily) and off the coast of northern Norway (called the Maelström and feared by the early Viking deep-sea voyagers). The action of the tides causes rapid changes in the littoral environment at the edge of the seas. In the course of the day there are recurring variations of water temperature, salinity, alkalinity and silicate and oxygen content. The animal and plant life of the littoral regions also experiences the effects of surf and storm. The tidal zone has resulted in some interesting adaptations of organisms to this unique marginal environment of the oceans.

Waves

Waves may be described as any movement of sea-water, ranging in magnitude from localized cat's-paw ripples that barely ruffle the surface to global tidal movements that traverse the entire Earth's surface. Waves are due to a number of forces such as wind, earthquakes and the Sun and Moon (tides). Waves can be explained simply as an advance of energy through the movement of water.

Wind Waves Characteristics

Waves caused by wind are due to the frictional drag of moving air over the sea. Mariners have a rule of thumb which relates wave height to wind velocity: it states that the height of a wave in feet ordinarily will not be greater than half the wind speed in miles per hour i.e. an 80-m.p.h. (130-kilometre-per-hour) hurricane produces 40-foot (12-metre) waves.* However, this is only a general guide, and many factors affect wave-height, -frequency and -length. The highest point of a wave is called the *crest* and the lowest point the *trough*. The *crest-length* of a wave is the distance along the crest of an individual wave. The *wave-length* is equal to the distance between two parallel crests (Fig. 7).

A more scientific rule for estimating wave characteristics is as

* The supposed world record for freak wave height is 34 metres (112 feet). One of the authors (P.L.B.) experienced waves over 25 metres (80 feet) high in the Southern Ocean near Kerguelen Island in 1953.

Table D The Beaufort Scale (the relationship between various wind velocities, the appearance of the sea and wave-heights)

Force	*Mean Wind Speed (knots)*	*Description*	*Sea*	*Height of Waves (feet)**
0	0	Calm	Sea like a mirror.	—
1	2	Light air	Ripples with the appearance of scales formed.	—
2	5	Light breeze	Small wavelets, more pronounced. Crests glassy appearance, do not break.	$\frac{1}{2}$
3	9	Gentle breeze	Large wavelets. Crests begin to break. Foam of glassy appearance. Occasional white horses.	2
4	13	Moderate breeze	Small waves, becoming longer; fairly frequent white horses.	$3\frac{1}{2}$
5	18	Fresh breeze	Moderate waves, more pronounced long form; many white horses. (Some spray.)	6
6	24	Strong breeze	Large waves, white foam crests more extensive everywhere. (Spray.)	$9\frac{1}{2}$
7	30	Near gale	Sea heaps up and white foam from breaking waves begins to be blown in streaks along the direction of the wind. (Some spindrift.)	$13\frac{1}{2}$
8	37	Gale	Moderately high waves of greater length; edges of crests break into spindrift. Foam blown in well-marked streaks.	18
9	44	Strong gale	High waves. Dense streaks of foam. Sea begins to roll. Spray may affect visibility.	23
10	52	Storm	Very high waves with long overhanging crests. Foam in great patches blown in dense white streaks. Surface of sea takes white appearance. Rolling of sea becomes heavy. Visibility affected.	29
11	60	Violent storm	Exceptionally high waves. (Small and medium-sized ships lost to view behind waves.) Sea is completely covered with long white patches of foam. Edges of wave-crests blown into froth. Visibility affected.	37
12	68	Hurricane †	Air filled with foam and spray. Sea completely white with driving spray; visibility seriously affected.	45 †

* Refers to open sea. Closer inshore waves will be steeper and not so high.

† To record the velocity in knots of Hurricanes of greater intensity than Force 12 the scale is extended: 13, 76 knots; 14, 85 knots; 15, 95 knots; 16, 105 knots; 17, 114 knots and over.

Note: to convert wave-height in feet to metres multiply by the factor 0·3.

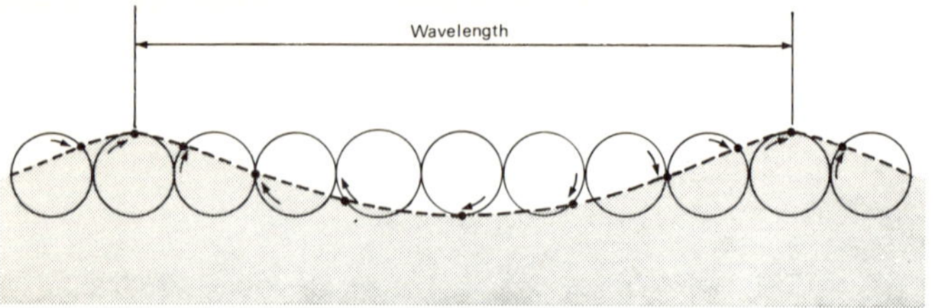

Fig. 7 Cross-section of an ocean wave travelling in a direction from left to right. Circles represent the orbits of water particles in the wave. At the surface (above) the orbital diameter described by a water particle is equal to the wave height; at a depth of half the wavelength, the orbit of a water particle is only one-twenty-fifth that of a particle at the surface.

follows: Wave-speed in knots = 3·1 times the period in seconds. Wave-length in feet = 5·1 times the square of the period in seconds.

Waves vary considerably in period.* Cat's-paw ripples have a period of only a fraction of a second, while huge ocean waves may have periods nearer to a quarter of a minute.†

A scientific wave theory was not thought out until the early 1800s when Franz Gerstner, a German scientist, showed that water particles in a wave motion moved in circular orbits, the diameter of which is equal to the height of the wave. Early observers had noted that passing waves did not transport objects horizontally, but rather back and forward and up and down; however, they could not think of the reason. The seventh Astronomer Royal, Sir George Airy (1801–92), was greatly interested in the theory of waves and later elaborated on Gerstner's earlier work.

A particle of water moving in an apparent circular orbit actually traces out a curve which approximates to a trochoid (Fig. 7).

The development of a wave depends on two factors, the force of the wind and the time and *fetch* (distance) over which the wind is free to work on the sea. In a limited space such as a bay or inlet, the waves produced on a stormy day will not have sufficient room to attain their maximum development. However, in the unrestricted

* Wave period is defined as the time taken for two wave-crests to pass a given point.

† For example, a wave with a period of 20 seconds has a speed of 62 knots (112 kilometres per hour), and its wave-length is 2,000 feet.

Southern Ocean, in the region of the Roaring Forties, waves always reach their greatest development.

Waves become complicated when subsidiary waves reinforce another set of waves. When the two wave-crests combine, a large wave results greater than the average wave of either source. But when one crest falls in another's trough, they tend to cancel each other out. The old sailors' yarn that every seventh wave is the largest is a myth, and the frequency of extra-large waves is very variable.*

Waves change shape when they move out from where the winds generated them. Their crests become lower and more rounded and tend to move in trains of similar height and period. When this occurs, it is usually called a swell, and sometimes a swell may be traced for thousands of miles as, for example, in the Southern Ocean.

During its development, if the height of a wave approaches one-seventh of the length, the angle of the crest is reduced to about 120° and the crest becomes unstable. It then moves faster than the base of the wave and topples forward to become a white cap or 'white horse'. Short waves reach maximum height more quickly and are then destroyed, while longer waves can accept more continuous energy from the wind and continue to increase in size.

When a wave approaches a shoreline and meets shallower water, it changes character. At a depth of water 1·3 times wave-height, the water particles have no room left to complete their cycle of circular motion, and the waves break to form surf. When air is trapped by a collapsing wave, it explodes with a roar. As a moving train of waves advances at only half the speed of its individual waves, the waves arriving on shore are only remote descendants of waves generated in the ocean.

Tsunamis

Tsunami is a Japanese name given to a wave-form which is the most destructive ocean wave known to man. Western oceanographers decided on the use of this Japanese name simply to avoid the term 'tidal wave' which had long been incorrectly used to describe this phenomenon. Ironically, however, it is now Japanese

* In storms or hurricanes wave patterns become numerous and highly complex. Over long fetches of open water the larger waves predominate, since a long wave can accept more energy from the wind and rise much higher than a short wave.

oceanographers who are embarrassed by its usage, since in Japanese the name does mean tidal wave!

But a tsunami is nothing to do with tidal phenomena and is caused directly by an earthquake or volcanic activity adjacent to, or under, the ocean. When an earthquake or submarine landslide occurs or a volcano erupts on the seabed, a *seismic wave* is created, and once transmitted to the water, it will travel great distances with very little loss of energy. Seismic waves triggered in the Pacific Ocean have been detected round the shores of Great Britain and other European countries.

Although the height of a tsunami in mid-ocean is only a few feet – so that a ship may traverse one without recognizing it – they create devastating effects near the shore. Tsunamis have periods of between 15 and 60 minutes and wave-lengths measuring several hundred kilometres, and they travel in the open sea at speeds in excess of 720 kilometres (450 miles) per hour. On reaching the shelving shoreline, the tsunami wave begins to 'feel' bottom and in doing so creates a great disturbance on the bottom sediment layers. The wave is braked to a speed of about 80 kilometres (50 miles) per hour, and this causes the water behind to pile up rapidly so as to form a colossal breaker, which may rise to a height of 60 metres (200 feet). As it cascades ashore, nothing can hold it back, and it spills inland for hundreds of metres, demolishing everything in its path, often carrying large vessels with it which it then strands like flotsam at great distances from the normal shoreline. As the wave subsides, forming an irresistible undertow, the seabed near the coast may run dry for hundreds of metres seawards.

One of the most famous tsunamis of the past was created by the Krakatoa eruption on 27 August 1883, when a cubic mile of volcanic cone was blown skywards, and the great hole left by the explosion was invaded by the sea. The inrush of the sea set up a disturbance of water with a period of about 2 hours. Nearly 40,000 people perished as mountainous waves travelled outwards at speeds between 560 and 720 kilometres (350 to 450 miles) per hour. These waves were detected in the English Channel 32½ hours later, fortunately in a very much modified form.

The Pacific Ocean* is particularly susceptible to tsunamis, and

* In the period 1946–1954 there were fifteen Pacific tsunamis, but fortunately only one of them was of first magnitude.

now an international early warning system is in operation. Seismic sea-wave detectors are positioned in key localities, and when a seismic wave event occurs, warning in advance of the first wave-front can be given to neighbouring areas (Fig. 8). Nevertheless, many people still perish, and over a span of centuries hundreds of thousands have died as a result of tsunami waves, and they represent one

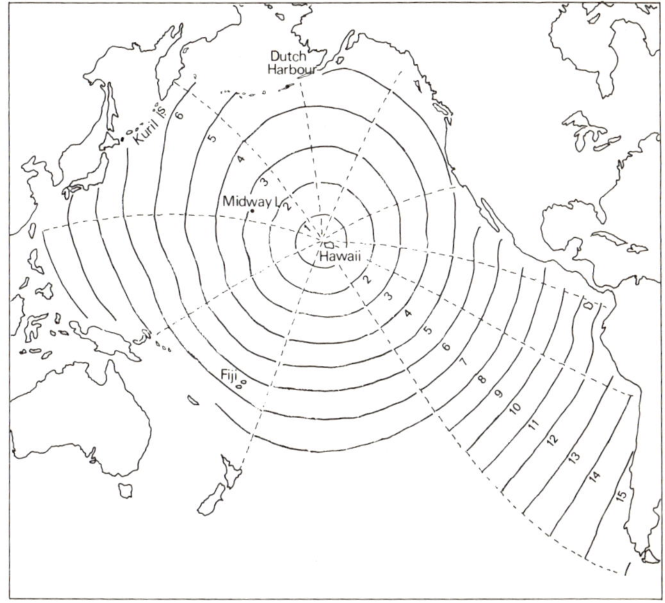

Fig. 8 Chart showing the time plot in hours for a seismic sea wave (a tsunami) to reach Hawaii from various parts of the Pacific Ocean. When the exact location of an earthquake epicentre is plotted, a warning can then be given of the probable time of arrival of the resulting seismic sea wave.

of the greatest natural disaster forces known to modern man. They are matched only by freak storm surges of water along low-lying coastlines, or when hurricanes, cyclones and tornadoes pile up water within the narrow confines of an area – such as occurred in the Bay of Bengal in the great disaster which befell the delta coastline of the mouth of the Ganges river in 1971.

P.L.B.

5 EVOLUTION OF THE OCEANS AND CONTINENTAL DRIFT

In the past the seas and oceans were battlegrounds for various geological ideas, for even the earliest geologists realized that the seas and oceans played an important role in shaping the Earth's surface. Before the nineteenth century, the biblical flood was accepted as *a priori* principle in all scientific debates as the chief agency for shaping the Earth. The Neptunists believed that all rocks on the Earth had formed in a great primeval ocean either by crystallization or by sedimentation. This view was opposed by the Vulcanists who believed that all rocks were originally volcanic, and present-day sedimentary rocks were simply their eroded remains.

Slowly the picture of the origin of different rocks emerged. William Smith, a British canal surveyor, studied the fossil fish beds in the Old Red Sandstone of England, and George Cuvier, the French naturalist, was able to show, from his geological explorations, that different fossils had origins in different environments. The early geologists, however, had little conception of the length of geological time, and to account for the noted great changes on the surface of the Earth, they thought it necessary to invoke the agency of sudden catastrophic events such as a deluge. One exception was James Hutton who believed that all changes could be accounted for by a long, slow, almost imperceptible process. Later Charles Lyell, following the example set by Hutton, adopted the phrase 'The present is the key to the past' and developed a great thesis that one could account for most of the present-day land forms by natural agencies such as wind, ice, running water and wave action. In this scheme of things, cataclysmic events such as earthquakes and volcanoes played only a minor role in shaping the Earth. In this way modern geology was founded on the principle known as uniformitarianism which dominated geological thought until the twentieth century. Lyell's book, setting out his great thesis in plain, easily understood language, was published in 1830 and entitled *Principles of Geology, Being an Attempt to Explain the Former Changes of the Earth's Surface by*

Reference to Causes Now in Operation. The title alone was the message of the thesis in a nutshell, and the book was adopted as the geological bible. Although nowadays geological thinking is passing through a great revolution, Lyell's catch phrase 'The present is the key to the past' still largely holds true.

Until quite recently many marine geologists believed that the present-day oceans and their basins had changed little since the Earth's crust solidified some 3–4,000 million years ago, in spite of tantalizing pieces of evidence to show that this could not be true. The trouble was that scientists in the recent past tended to work in isolated disciplines of thought. Demarcations in the various branches of science were clear-cut and rigidly adhered to. Geology in particular was a moribund science. Many scientists truly believed that most of the major principles had already been enunciated in the nineteenth century by people such as Lyell, and all that remained was to fill in the mundane and minor details.

After World War II, the scientific outlook began to change, and many of the old dogmas were challenged. Scientists are now quite used to the idea of utilizing (or lifting) ideas from other sciences in order to rethink and solve their own particular problems. It was the advent of the International Geophysical Year in 1957 (IGY) which gave both marine geology and oceanography a new lease of life so that today they are among the most exciting and developing sciences of the 1970s.

* * *

Although the change about of ideas concerning the oceans and continents did not really get under way until the late 1950s and the early 1960s, Alfred Wegener, an obscure German meteorologist, was the first man in modern times to put forward the revolutionary ideas of continental drift. In 1915, Wegener announced his theory that during geological time the large continents had drifted apart from a primitive super-continent. But Wegener was not the first one to put forward this idea. Francis Bacon hinted at it in 1620 in his book *Novum Organum*, and in 1801 Humboldt actually suggested that South America and Africa may have drifted apart forming the intervening Atlantic Ocean, which was probably only a sea-filled valley. No one who looks at a map or a globe can help noticing the apparent fit of the two continents. However, most nineteenth-century scientists believed it to be a matter of pure coincidence.

Others could not ignore the evidence. In 1858 a book appeared written by Antonio Snider-Pellegrini entitled *Creation and its Mysteries Revealed*, which put forward the case that the American continent had become separated owing to an earth-shattering catastrophy such as the great flood of Noah. The book was written much

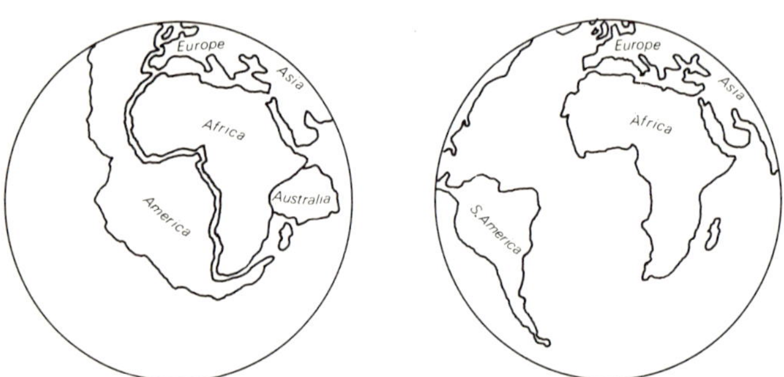

Fig. 9 World maps published by Antonio Snider-Pellegrini in 1858 to illustrate his ideas about continental drift and the similarity of Carboniferous fossils in Europe and the Americas.

in the vein of other 'catastrophe' books like Whiston's *New Theory of the Earth* (published 1690), which utilized the biblical deluge to describe all manner of earthly changes without a shred of hard 'scientific' evidence to prop up the ideas. Snider-Pellegrini, however, provided evidence by citing similarities in Carboniferous fossils in North America and Europe. His book also shows the earliest extant map of the relative configuration of the continental masses before separation is supposed to have taken place.

Later in the nineteenth century the Austrian geologist Edward Suess coined the word Gondwanaland which, it is supposed, represented the original land mass in the Southern Hemisphere from which present-day South America, South Africa, India, Australia and Antarctica broke away* (see Fig. 29).

* A supposed super-continent to the north was named Laurasia. The problem has long been argued whether in primeval times either a single or two super-continents were formed, but as yet there is no agreement about this.

What Wegener did in 1912 was to try and develop the earlier speculations and transform them from the realms of a Jules Verne romance into respectable scientific theory supported by reasonable argument. He argued in his book *The Origin of Continents and Oceans* (published in 1915) the premise that if huge continents had once cracked into slabs which then separated, then pieces would be expected to contain telltale evidence of their previous attachments. For example, zones which had been shattered by ancient earthquakes or radically altered by mountain building could be matched by a careful inspection of the present-day continental rafts. The animal kingdom should also provide evidence, for a common stock would begin divergent evolutionary trends immediately after separation took place.

Wegner in his drift theory showed how the present-day configuration of continents and oceans came about in successive steps beginning about 270 million years ago. He supposed his original land mass to be a single continent he called Pangaea and a single great ocean known as Panthalassa, which covered the rest of the globe. Superimposed over the original land mass were the secondary shallow seas known from their fossil records.

Wegener's principal evidence purported to show that Greenland and Europe had separated at a rate of about nine metres a year during the period 1870 to 1907. When his critics did some quick mental arithmetic, they called attention to the astounding fact that such a rate of change would be equivalent to the circumnavigation of the Earth by a 'wandering' continental block in two million years! This was a ridiculous figure, and his geologist critics smiled condescendingly, remarking that this was the direct result of a meteorologist entering into matters beyond his experience. When the so-called drift was independently remeasured and found to be wanting, this reinforced the initial reception to the idea. The classical evidence provided by the apparent fit of South America and Africa surprisingly served Wegener no better. But perhaps this was exactly where his knowledge and experience were lacking. Had he been an oceanographer, he might have thought of looking beneath the sea for his answers as well as above it. Although it is an established fact that the continental sea edges do appear to fit, at best this is only a generalization, and as far as the coastlines are concerned, it does not provide a truly conclusive argument. Had Wegener instead used the boundaries of the continents as defined by their 600-fathom-line continental shelves, it might well have given rise to a modern cry of 'Eureka!' (as happened in the 1960s when, with the assistance of a computerized reassembly, all further doubts were cast aside). Wegener's theory was also damned by the fact that he was not able to cite the forces inside the Earth to account for his drifting continents. As a result, Wegener's theory was banished to lie among the hundreds of other crank theories that unfortunately abound in science, and his book was received with derision and contempt.*

In science a new theory is only accepted as a working hypothesis as long as it is successful. All theories are working hypotheses, for no one can say that a particular theory is absolutely correct, since we have no way of knowing. What science really does is to narrow the field of speculation. In this context one cannot really blame Wegener's contemporaries for being critical, since none of Wegener's ideas appeared to fit the known facts. In hindsight, however, we known that Wegener's ideas are basically correct, but this is only because new evidence has come to light. For example, the survey techniques for measuring the supposed drift between Greenland and Europe were too crude in Wegener's day, the circumstances of the continental fit have already been cited, and it was not for another half century that someone did come up with a model to provide a logical reason for drift to occur. It is only too easy to look back in hindsight and state that Wegener's dossier of evidence – in spite of its shortcomings – should have been sufficient to convince everyone in 1915. Yet in the 1950s Fred Hoyle, considered a radical, unprejudiced thinker, thought continental drift an unlikely phenomenon, and even today there are a handful of critics who still remain unconvinced and offer other ideas to explain away the now overwhelming evidence.

What rescued Wegener's theory of continental drift from oblivion was the new science of geophysics which, as it developed, intruded deeply into geology and oceanography. Methods and ideas not previously available were introduced – one scientist likened it to a bronze age overtaking a stone age.

* Poor Wegener was ill-fated and later died tragically of starvation on an expedition to Greenland in 1930.

A consequence of Wegener's drift theory would be wandering poles. The magnetic poles are at present located very approximately in the direction of the geographical poles. Even from day-to-day experience the magnetic poles have been found not to be fixed points but constantly to swing back and forth, and over a period of about 500 years they describe small circles. Over the geological time-scale, however, the magnetic poles appear to wander haphazardly across the entire Earth's surface; at times they have been located 90° away from the present-day geographical poles of rotation.

The generation of the Earth's magnetic field and continental drift have a close connection. The Earth's magnetic field is probably due to convection currents in the fluid core which act like a hydromagnetic self-exciting dynamo and produce an electromagnetic force. During the course of the Earth's 4,600-million-year history as a planet, it has switched polarity several times* (see Fig. 11) so that the South Pole became the North and the North the South. These reversals can be studied through the modern science of paleomagnetism,† which concerns itself with the measurement of weak fossil magnetism remaining in certain rocks. Some rocks have been found which show evidence of being magnetic at a period when the Earth's magnetic field was in the very act of switching polarity. These magnetic reversals have had far-reaching consequences for the continental drift theory.

During the world cruise by the research vessel H.M.S. *Challenger* in 1872–5, a survey was made on the bed of the mid-Atlantic which showed a submerged mountainous range – what was then thought to be a sunken geological feature. Later oceanographic expeditions made similar soundings in both the North and South Atlantic, and by the 1950s these soundings were shown to represent a huge continuous ridge-like feature running the entire length of the ocean. The other large oceans were also found to have similar ridges. Geologists and oceanographers soon realized that the world-wide pattern of underwater ridges was quite different from those found on the continental masses. Rock samples from these ridges are all basaltic in

* The reason for this is at present obscure.

† Paleomagnetism began as a science in the 1890s in Italy, through the study of large Greek Attic vases made by the Greeks in the sixth century B.C. These vases are faintly magnetized and retain previous directions of the Earth's magnetic field.

origin, indicating unmistakable volcanic origins. These rocks are similar in composition to the rocks of numerous volcanic islands scattered far and wide over the Earth's oceans.

When these ridges were further explored, they were shown to be areas of abnormally high heat flows, thus adding evidence to the volcanic hypothesis. The question was: What significance do the ridges have in the evolutionary processes in the oceans and coastal rocks? Many geologists believed that all the folded and faulted movement zones, including the mountain ridges of the Earth, were caused by the Earth's contracting as it grew older. Isaac Newton was the first to suggest that the Earth was shrinking. There is an apocryphal story which tells that Newton, having observed the famous apple fall, and received his ideas about gravity, took the fallen fruit indoors to remind himself of his inspiration. After a few days he noticed that the apple skin had shrivelled, and this supposedly inspired him again to the analogous idea of a shrivelling and shrinking Earth's crust! Were then the mid-ocean ridges caused by a shrinkage of the crust? If they were, why the high flow of heat emanating from them?

But advocates of the Earth-shrinking hypothesis were thrown off balance when new evidence was provided through radioactive dating. Some of the rock samples dredged up from the ridges showed that they were relatively young rocks, certainly not more than 10 to 12 million years old. These might have been squeezed out of the Earth by the shrinking crust on either side of the ridge. Nevertheless, to many geologists these rocks suggested – instead of a contracting Earth – an expanding one, and the ridges were the cracks in the eggshell where the Earth's crust has fractured to allow the expansion to take place.

Historically, the idea of the possibilities of the Earth's expanding dates to the 1930s. It was then that physicists began to speculate about the gravitational constant – the force *G*. They posed the interesting question: Had *G* varied over geological and astronomical time? Any variation in *G* would have far-reaching consequences for the Earth both as an astronomical body and during its geological evolution. By jiggling around with the force *G*, theorists are able to explain (rather glibly) all manner of cosmic and earthly events which seem to defy rational explanation if *G* is assumed to be a fixed constant.

Whether the Earth, as a whole, has in fact greatly expanded over geological time is a question still fiercely debated. However, the arguments need not worry us here, for it is now beyond all reasonable doubt that the mid-ocean ridges are the sources of upwelling from inside the Earth which bring about a continuous change on the bed of the oceans. This upward movement at the site of the mid-ocean ridges is brought about by convection cells operating within the mantle of the Earth. As the material arrives on the surface, it divides, and in the case of the Mid-Atlantic Ridge, carpets of material spread east and west. But if material is brought constantly to the surface, what happens to the ocean floor pushed outwards on either side of the ridge?

The ocean deeps provided the clue here. Exploration of the deeps using modern geophysical techniques showed that gravity was less than it should be. The only explanation of this could be that the deeps, or trenches, were the result of deformation in the ocean floor caused by material disappearing downwards back into the mantle to complete the gyre in the convection current circulation loop.

Thus the picture was revealed of an ocean floor constantly being replenished at one side and disappearing at the other – just like a slow-moving conveyor belt (Fig. 10). Later crack-like valleys were found running down the median lines of all the ridges, and this is where the new material emerges from below the surface. Most of the jigsaw pieces now fitted into place. One of the greatest discrepancies

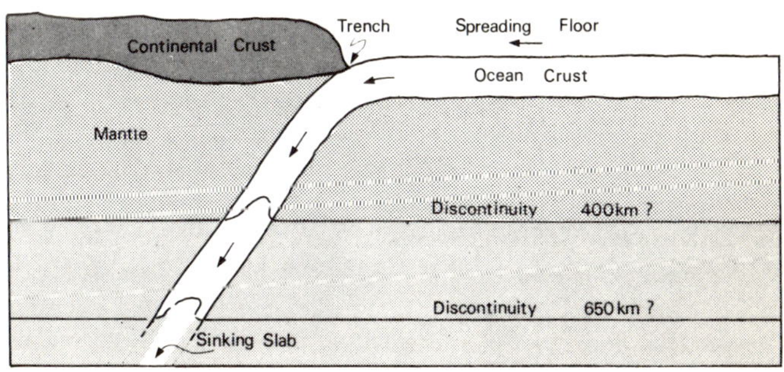

Fig. 10 A schematic representation showing the spreading ocean floor descending into the mantle where it meets the continental crust and forms a trench (or deep).

in the classical model of permanent ocean basins was in the thickness of soft sediments lying on the ocean bed (see p. 66). If the oceans were of great age, the sediments should be enormously thick. The sedimentation rate is easily measured, and the measured rate gave results quite out of accord with ideas of an unchanging ocean. The red clay deposition in the oceans occurs at a rate of 0·9 centimetres (0·35 inches) per 1,000 years, and the globigerina ooze at 1·2 centimetres (0·47 inches) per 1,000 years. If the seas of today had begun in Cambrian times (about 500 million years ago), then the sediments would be at least 5 kilometres (3 miles) thick. Nowhere in the world have sediments been measured which are greater than 300 metres (1,000 feet). The sedimental evidence should have pointed the finger of suspicion long before it did. However, wrong measurements were partly to blame, and in the older literature one comes across sediment thicknesses quoted in many thousands of feet.

But the most revealing evidence of the replenishment of the ocean floors harks back to the previously mentioned evidence provided by the Earth's magnetic properties and in particular its inexplicable ability periodically to reverse the direction of its magnetic field. Ocean research vessels, towing behind them magnetometers, began routinely to measure the Earth's magnetic field over some peculiar magnetic anomalies, which geophysicists had located along the seabed. In the Pacific it was noted that the anomalies appeared as stripe-like features which were interrupted by great faults on the seabed, but that they ran parallel with the general line of the ocean ridges. The pattern appeared as a chaotic jumble beyond comprehension, until someone had the idea of using the magnetic reversals as an analytical tool. This proved to be the Rosetta Stone in translating the story of seabed evolution. By matching up the magnetic lineations, including the reversals, and assuming that the spreading rocks moved symetrically outwards, each of the catalogued magnetic epochs (see Fig. 11) was identified on either side of the ridge.

The changes in the Earth's magnetic field take place at widely varying intervals. Occasionally about a million years go by without changes, and during the Permian period there were apparently none recorded for at least 20 million years. But the magnetic message is indelibly branded on the new material as it emerges from the ridges and can never be eradicated. In the Pacific Ocean some subsequent

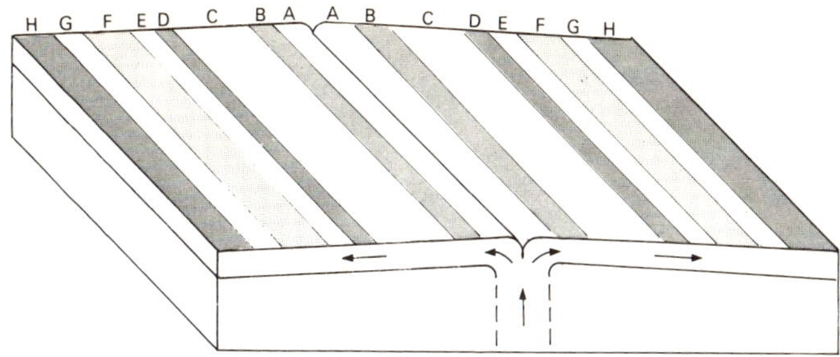

Fig. 11 A schematic block of ocean floor showing mantle material extruded at a ridge. As new rock emerges and spreads laterally either side, it adopts the prevailing magnetic polarity of the Earth. At irregular intervals the Earth's magnetic field reverses polarity, and these reversals are recorded permanently in the sea-floor rocks as a series of matched alternate bands, or stripes. The white stripes represent normal polarity, and the shaded stripes reversed polarity. The youngest extruded rock, A–A, is nearest the ridge; H–H represents the oldest rock (see also Figs 30 and 31).

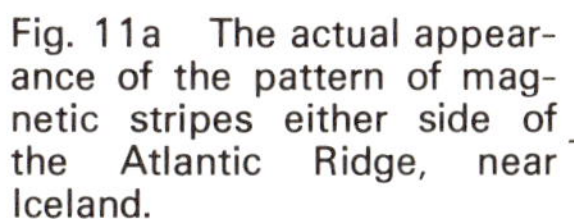

Fig. 11a The actual appearance of the pattern of magnetic stripes either side of the Atlantic Ridge, near Iceland.

faulting of the seabed has displaced parts of a single magnetic stripe a distance of over 1,120 kilometres (700 miles) (Fig. 30).

Other evidence of sea-floor spread (and of rate of spread) is provided by basaltic samples dredged up from the Mid-Atlantic Ridge. These were aged by using a 'fossil' fission track technique. All rock fragments contain traces of decaying uranium-238 which, in decaying, leaves telltale 'fossil' tracks in the sample that can be etched out and then counted under a microscope. The track density is directly related to the time for which the isotope has been undergoing nuclear fission. Among four rock samples dredged up, one came from the median valley bisecting the ridge which provided an age of 13,000 years; another 6·5 kilometres (4 miles) out from the centre gave 290,000 years; the third, 16 kilometres (10 miles) out from the ridge, 740,000 years; and the fourth, 58 kilometres (36 miles) out gave 8 million years. Taken together these samples show that the sea-floor spread at this particular spot is at the rate of one inch per year.

The chronological record of sea-floor spread can be related to island geography. If the floor of the ocean is moving 'sideways' – either side of a parent ridge like conveyor belts – then all the volcanic islands arising from the ocean floor should be carried along with it at the same rate. Although volcanic islands now appear to have a random distribution, they all originated in the median valley of the mid-ridges, so that their present distance away from the ridge is a direct measure of an island's age. The younger islands should be nearer the ridge than the older ones. This is found to be exactly the case. Likewise, the bottom sediments should show a similar chronological distribution. A special magnetometer can identify the magnetic reversal pattern in the sediments lying above the igneous basement rocks on the seabed. Within the sediments are minute particles of iron which at the time of deposition aligned themselves along the Earth's magnetic field. Those deposited in high latitudes show a pronounced dip as would a compass needle. Cores taken through the sediment layers show the chronology along a vertical scale, so that we have a time-scale in two directions which may be closely correlated.

The rates of sea-floor spread have now been measured with great precision. The slowest rate appears to be from the Mid-Atlantic Ridge and the Carlsberg Ridge in the north-west region of the Indian Ocean where the new floor emerges at the rate of 1 centimetre (0·4 inches) per year, either side. The fastest rate is located on the ridge of the East Pacific where the rate is 8 centimetres (3·2 inches) per year, either side, or a total of 16 centimetres (6·4 inches) adding both together, which is the true rate of spread of new ocean floor. Using the above rate the ages of various oceans can be calculated by reverting to simple arithmetic, and the entire seabed of the Pacific can be accounted for in 100 million years.

When the mid-ocean ridges were even more critically examined, they were found to be highly fractured zones. The magnetic stripe anomalies had shown this (see above), and in some areas the displacements, called *transform faults*, are of considerable magnitude. The recognition of these large displacements brought about a development of the theory of sea-floor spread and continental drift

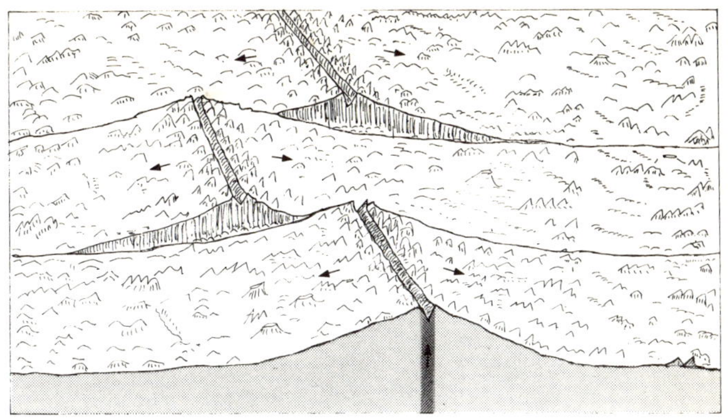

Fig. 12 Transform faulting along a sea-floor ridge which gives rise to lateral displacements of large sections of ocean floor. Transform faulting is revealed by the distinctive magnetic bands, or stripes, impressed in extruded rocks at the time they emerge at the ridge from the mantle. When samples of these rocks are dated, the chronological displacement patterns are readily discernible along either side of a ridge (see Figs 11 and 30).

even further. Simple ocean-floor spreading either side of a ridge raises some difficulties. For example, in the Pacific there are numerous deep trenches (see Fig. 22) to provide dustbins, or trash cans, where old ocean floor descends back into the mantle, but in the Atlantic there are no such trenches, apart from near the West Indies. It was apparent that the theory needed drastic modification if it were to remain acceptable.

In the end it was the Pacific Ocean which provided a significant clue: in the East Pacific a trench running up the west coast of the Americas merges with the land near the coast of California and appears to join up with the famous San Andreas fault, whose sudden movement was the direct cause of the San Francisco earthquake disaster of 1906. The questions now raised were: What influence did ocean-floor spreading have on continental masses? And since the Pacific floor and other ocean floors had been shown to be very young features on a geological time-scale, what had the surface of the Earth been like before continental drift began perhaps some 100 million years ago? Now geologists posed the real question: Was ocean-floor spread the result of the continental drift, or was continental drift a consequence of sea-floor spread? If one weighed the evidence – particularly that provided by the indelible magnetic pattern impressed in the ocean floors – there could only be one solution: the oceans and the continents were broken up into large independent sections and moved adjacent to one another as rigid plates of material which disregard the natural physical divisions between oceans and continents. A simple explanation of continental drift was possible if one assumed that the Earth's surface is made up of six major slabs of crustal material, or *plates*, as the geologists have termed them, whose demarcation lines follow the global pattern of high earthquake and volcano frequency zones, represented in the oceans by the great ridge system (see Fig. 31).

The Indian plate carries most of India and Australia plus all the north-eastern Indian Ocean, bordered in the south by the Indian Ocean Ridge and in the north by the Himalayas. The African plate takes in the whole of the African continent, plus the Atlantic Ocean east of the Mid-Atlantic Ridge. To the north it is bordered by the Mediterranean and to the north-east by the Red Sea. The Pacific plate is almost wholly oceanic, except for parts of California, and meets the North American plate at the junction of the San Andreas Fault line.

By invoking the movements of the plates against one another, the whole idea of continental drift, sea-floor spread and mountain building are unified into one beautiful all-embracing theory. By a detail analysis of *plate tectonics*, which is the name given to this phenomenon, we can look into the past geological history of the world and work out how the present-day configurations of the Earth's surface came into being. In the case of the Indian plate where it meets the Asian plate, the Himalayas resulted. It is likely that before the two plates collided, there existed a deep ocean trench between them. As the Indian plate advanced, this trench was consumed and now lies buried under the rocks of the Himalayan Mountains, which once formed the bottom of an ancient sea which geologists call the Tethys Sea.*

From an examination of the geomorphology of the plate abut-

* Tethys in legend was one of the Titans and the wife of Oceanus.

ments, it is possible to work out some simple rules of what actually occurs when two plates brush or collide. For example, when an oceanic plate meets a continental plate, e.g. where the Pacific plate meets the North American plate, the oceanic plate is pushed below,

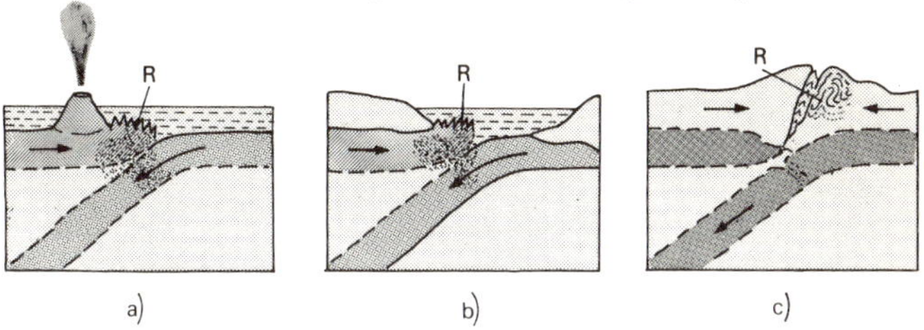

Fig. 13 Plate boundaries showing various situations when two plates meet. a. Two plates of oceanic crust are moving in opposite directions. One plate has underthrust the other, and an arc of volcanic islands has erupted on the upper plate. The downward-moving oceanic plate has lost some of its material (R) to the leading edge of the other plate. b. oceanic plate is pushed down under a continental margin, and another looming continent is being carried inexorably along above it. c. Two continents have collided and formed new mountains, parts of which contain relics (R) of the now vanished ocean floor. In this situation the leading section of the underthrust plate may break off below the new mountain range, and the remaining material may then continue to fall back into the mantle.

or underthrust, resulting in a gigantic crustal buckling that gives rise to the formation of a range of mountains of the *cordillera* type.* When two continental plates abut, the result is a belt of intensely deformed rocks such as the Himalayas, where neither plate gives way, so that the consumed material from either plate is buckled into a chaotic system of broken folds of great complexity. The picture at the site of the Western Pacific deeps resembles that first conceived by sea-floor spreading. The Pacific Ocean plate is pushed downwards (Fig. 13a) into the deeps frontiering the volcanic island arcs. The arcs themselves are subject to volcanic activity owing to the tectonic movements below and alongside. As the ocean crust begins to descend back into the mantle, there is a partial melting of the material, some of which reascends as liquid magma and re-emerges on the Earth's surface as familiar volcanic lava.

Along the western seaboard of the United States we see the

* Parallel ranges of mountains.

situation where the continental Americas plate is over-riding the Pacific oceanic plate. The result has been that a deep trench which existed previously in early geological times, when the Americas were farther east, is completely obliterated. At the present time a large part of California is moving northwards. Part of California is attached to the Pacific plate and part to the over-riding continental mass, giving rise to the famous fracture zone, the San Andreas fault.

In the region of the Red Sea we have an embryonic ocean in the making (see p. 19). The Red Sea contains some unusual deep pools of hot metalliferous brines and thick deposits of metal-rich sediments. One brine pool, located by the British research vessel H.M.S. *Discovery* in 1964, has a temperature of 44° C. (111° F.) and is about 13 kilometres (8 miles) square; this is now known as the *Discovery* Deep. Another pool known as the *Atlantis II* Deep – discovered by the U.S. research ship of that name from the Woods Hole Institute – is even larger and has a measured temperature of 56° C. (133° F.). These large areas of brine exist as thermally isolated pools of super-saturated water with a salinity of about 290 parts per thousand, as against average surface water of 38–40 parts per thousand.

One of the sediment specimens recovered by the *Atlantis* looked and smelt like scalding black tar. Many of the sediment specimens contain remarkably high metal concentrations. This discovery has caused geologists to think again about the whole subject of mineral ore deposition. Existing theories may be quite wrong; at the present time new facts are coming to light faster than the scientific journals can publish them! It seems certain that hot oceanic brines play a major role in world-wide mineral deposition.

From an oceanographic point of view the significance of the hot brine sinks is that they lie on the continuation of the line of the mid-ocean ridge extending from the north-west part of the Indian Ocean. In the far north of Ethiopia lies a dry-land area called the Afar Triangle. This triangle is important in the reconstruction of the Red Sea basin before it began to open up 20 million years ago, for the south-western corner of Arabia fits into it like a missing piece of jigsaw puzzle. It has been shown from local field work that this now dry-land triangular area is undoubtedly a recently formed part of the seabed which is temporarily raised above sea-level. The area contains a remarkable flat-topped volcano, Mount Asmara, which consists of shards of volcanic glass such as one would expect had the feature been formed under water. Mount Asmara is also remarkably reminiscent of the flat-topped, under-water guyots, whose plateau-like summits are usually attributed to wave erosion (see p. 16).

Through the mechanism of sea-floor spread and tectonic plate movements, all manner of other outstanding scientific questions can be settled. It has far-reaching consequences to zoologists in explaining the distribution of life forms throughout the Earth, and the old ideas of doubtful continental bridges, by which animals migrated, can be forgotten.* The action of the great subcrustal machine – as sea-floor spread is colloquially known – can also explain the source of water now contained in the world's oceans. Much of the deep underlying rocks of the ocean floor consist mainly of serpentinized peridotite. At a temperature of 500° C., serpentine decomposes and yields juvenile primordial water. As new rock emerges at high temperature, the water sweated out is absorbed into the near-by ocean. Some inland mountain ranges, now locked up in continental masses, certainly were thrust up when two plates collided – such as the Urals that once formed the bed of an ancient ocean that divided Siberia from western Russia.

In Wegener's original concept, continental drift supposedly began 270 million years ago, but on the evidence of plate tectonics it seems more likely that such global movements have been operational for perhaps 3–4,000 million years beginning at the end of the 'lost' 500 million years in the early history of the Earth. Ocean bottom rejuvenation and the jostling of the major plates causes a complete transformation of all the surface features of the Earth, excepting the ancient shield areas which provide cores for the continental masses. At the present rate of sea-floor spread this means that the global oceans have renewed themselves at least fifteen times. Recent-day plate movements are causing the Pacific Ocean to close up, and the Red Sea to open up. On present available evidence it seems that the oldest piece of ocean floor lies to the west of Australia, and this has been estimated as about 225 million years old.

One of the direct economic consequences from the study of

* Except perhaps in Europe, South East Asia and the Bering Strait where land bridges were brought about by the last ice age.

continental drift is an idea that certain kinds of characteristic plate movements and interactions indicate sites of likely mineral deposits. In the future these might perhaps allow geophysicists to pinpoint areas rich in minerals – particularly areas rich in hydrocarbon fossil fuels. Some of these characteristic plate movements supposedly shaped southern California and north-west Siberia, areas both rich in hydrocarbons. Already several new areas have been listed which at present, in terms of crustal mineral resources, are largely unexplored. Such areas include the Labrador Sea, the Newfoundland continental shelf, northern Greenland, the Canadian Arctic islands, several areas round Australia, and the south-west and west Pacific. Thus at present it is very likely that there are major mineral resources lying undetected by traditional prospecting techniques.

P.L.B.

6 THE ENVIRONMENTS

Marine Deposits

The sediment material which immediately overlies the 'solid' seabed shows characteristic differences according to its geographical position, origin and environment. Bacteria contained in marine sediments, although much less abundant than in terrestrial soils, play a major role in replenishing the oxygen of the atmosphere. They are able to convert carbon to carbon dioxide that can then be re-utilized by phytoplankton which in turn releases free oxygen.

Littoral deposits are those found on beaches or in any shallow water between the high and low tide marks.

Continental shelf deposits include shallow water deposits found round oceanic islands. They are also sometimes referred to as *neritic deposits* after the mythical Greek god Nereus, a son of Oceanus, whose kingdom was the shallow sea. The term neritic was once confined to shallow water deposits of organic origin, and the two kinds of deposit are now often distinguished by the use of the description *organic* neritic deposit when applicable.

Deep-sea deposits are subdivided according to the depth at which they occur. The intermediate region is termed the *bathyal zone*, and the deep sea bottom the *abyssal zone*. The environment of the deep ocean trenches, or deeps, is called the *hadal zone*.

Marine deposits are also grouped according to their origin.

Terrigenous deposits are those derived from land and reach the sea through the action of rivers, ice, wind or by the erosive process of moving sea-water at its margins, i.e. waves.

Biochemical and chemical deposits are formed wholly or partly from sea-water which may incur involvement with organisms.

Organic deposits are formed from the accumulation of debris of marine organisms, in particular calcareous and siliceous shells.

Distribution of Marine Deposits

Present-day marine deposits show a world-wide distribution of each of the principal sources. Terrigenous deposits of shingle, gravel, sand and mud are found in the littoral, continental shelf and abyssal zones. Biochemical and chemical deposits are represented in the littoral and continental shelf zone by oolite sands, calcareous muds and evaporates; in the bathyal zone by glauconite and pyrites; in the abyssal zone by manganese nodules which 'grow' round nuclei such as shark teeth and whale earbones. Organic deposits are of two kinds, neritic and pelagic (from the Greek *pelagos*, open sea). The neritic are chiefly remains of the benthos (bottom dwellers) such as corals, sea urchins, molluscs, seaweeds, etc., which are formed in the littoral zone and round the continental shelves, and in the bathyal zone by coral muds. The pelagic deposits are derived from plankton and from the oozes of the bathyal and abyssal zones, such as diatom ooze, globigerina ooze, pteropod ooze and radiolaria ooze.

Red Clay Deposits

One of the primary deposits of the deep oceans, representing over 104 million square kilometres (40 million square miles) of seabed, is the red clay which owes its 'red' colour to the presence of iron oxide. But like many generalized names, red clay is not always red but rather a brownish colour, although it is quite red in isolated localities. The red clay is an hydrated silicate of alumina which plays an important chemical role in buffering the composition of sea-water, and it assists in controlling the slight alkalinity of sea-water at about pH8. The clay also holds a great deal of exchangeable silica, which it periodically gives up to maintain an oceanic balance. This control apparently operates over a geological time-scale, whereas short-term changes appear to be controlled by geological processes.

The origin of the red clay has been subject to much debate. It includes volcanic, aeolian and cosmic dust, plus solid debris dropped from melting icebergs and insoluble organic relics such as shark and cachelot teeth. It was once thought that much of the clay was formed from volcanic dust or was carried into the sea by river action, but now it seems that a large proportion of it is aeolian, picked up over deserts and carried into the upper atmosphere and transported long distances in high-altitude jet streams before eventually raining out.

Marine Deposit Ages

The deep sediments provide oceanography with a method of dating marine deposits. Measured rates of deposition (or sedimentation) vary from region to region, but in the Pacific and Atlantic the figure of red clay deposition varies between 0·07 and 0·2 centimetres per 1,000 years, and for globigerina ooze reaches up to 1·6 centimetres per 1,000 years. The thickness of the abyssal deposits, measured by seismic exploration, is about 300 metres (1,000 feet). In the late 1960s and early '70s, the U.S. drilling ship *Glomar Challenger* (see Fig. 34) has attempted to penetrate the full depth of the sediments, but each time has met with a thick layer of impenetrable chert, whose origin is probably due to the chemical changes in the oldest layers of silica-based animal debris.*

The total thickness of marine sediments, as a function of time, is quite small in relation to the geological time-scale, and at its greatest thickness represents a period of 300 million years. This provided an important clue in the formulation of new ideas about the age and origin of the oceans (see p. 56). If the Pacific basin were as old as some geologists once believed, the sediments should measure over 3 kilometres in thickness.

* The extensive chalk beds of England show similar bands of hard flint.

Marine Environments

Each of the environments in the oceans has its specialized population of marine organisms and life, which through natural selective processes has become highly adapted to its particular niche.* Below the surface of the oceans, temperatures fall quickly and form a thermocline or sharp boundary of temperature and density change (Fig. 54). Some ocean life has adapted to general conditions and is able to roam over a wide territory of varying depths and temperatures. Wide stretches of ocean are sometimes biological deserts, particularly great stretches of the Pacific upper levels. Phytoplankton which begin at the base of the ocean food chain can only thrive where the upwelling of nutrients is able to support conditions for photosynthesis to take place (see p. 204).

The *euphotic zone* (top layer) is the key environment for marine life in the oceans, for it is here that phytoplankton multiply to form the base of the great food-chain pyramid (see p. 210). The phytoplankton blooms in the spring in the temperate zone, in summer in high polar latitudes and all the year round in the upwelling areas where the steady offshore winds push the surface water to cause the upwelling of the all-important bottom nutrients. The herbivorous zooplankton (see p. 211) such as copepods, medusae and krill feed on phytoplankton, which in turn then represent a source in the chain for other nektonic (free swimming) creatures that inhabit the euphotic zone such as the baleen whales, sharks, seals, flying-fish, dolphins, porpoises, tuna, mackerel, small squid, sea-birds and the herring-like fishes. The herring-like fishes in particular feed on phytoplankton, which, being the most abundant food source in the ocean, allows the herring fishes to be one of the commonest and most abundant sources of fish in the world.

The euphotic zone, on the basis of its world-wide area, shows a

* Some of the fish that reside in Antarctic waters have adapted so as to survive in water two degrees centigrade below freezing point. Serum obtained from these fish contains chemicals of the hydroxyl groups similar to chemicals which are used by man in antifreeze agents in car radiators. Thus it would seem that through the process of natural selection the fish have evolved so as to manufacture their own antifreeze agents to allow them to reside in one of the most forbiddingly cold environments of the oceans.

whole range in sea temperatures. Some of its animal population is highly adaptable to these variations. For example, the large baleen whales like the blue, fin and humpback whales (Fig. 61) are equally at home in cold or warm water environments. However, their adaptation to different environments has a special purpose. These large baleen whales feed in cold water environments – in particular in the region of the nutrient-rich waters of the Antarctic Convergence (see Fig. 26) where the krill (*Euphausia superba*) are supported in huge populations by ideal environmental conditions. Here whales do all their feeding in summer, and on the approach of winter they migrate northwards to breed in subtropical regions. The southern humpback groups in particular show a remarkable migration pattern. Each group has its own breeding area to which it unerringly returns northwards – at some genetic command – at the onset of the southern winter. There are separate population groups off the west coast of South America, the west and east coasts of South Africa, and the west and east coasts of Australia, and one in New Zealand waters. During the 1950s and early '60s, when all these populations were hunted in both summer and winter, the southern humpback populations were almost completely wiped out. It has been estimated that the Australian groups will take at least fifty years to recover their stocks to make the humpback fishery worthwhile again.*

The *mesopelagic zone* (middle layer) is characterized by marine species such as lantern fish, hatchet fish, octopuses, large squid, angler fish, some varieties of shark, scarlet prawns and varieties of zooplankton. Among the large creatures which live partially in this zone is the sperm whale (Fig. 35) which descends to great depths to feed on giant squid. Sperm whales frequent depths up to 1,200 metres (4,000 feet) as is evidenced by carcasses which have become trapped in seabed telegraph cables. A number of sperm whales have been found caught up in seabed telegraph cables, and the probable reason is that they 'plough' the bottom with their lower jaw agape and attack a cable, perhaps thinking it to be the quiescent tentacle of a squid! Sperm whales are particularly well adapted to high under-

* It was not until the southern humpbacks were practically exterminated that the International Whaling Commission brought about a total ban on hunting southern humpback stocks in 1963.

water pressures and may remain submerged for periods longer than an hour. These whales are able to cope with the problems of narcosis without apparent ill-effects, and various theories have been put forward to account for their extraordinary quick adaptation to changing environments. Whatever the reason is, it seems likely that this ability is closely linked with the biochemical nature of their blubber fat.

The giant squid (the largest of the invertebrate animals), which provide the principal source of diet for the large-toothed whales, are less well known. Specimens have been recorded 18 metres (60 feet) in length, and many believe that even larger specimens inhabit the lower regions: indeed, many of the legendary sea monsters may be represented by these mesopelagic zone creatures, which have given rise to the legendary stories of Kraken off Newfoundland. The sighting reports of giant squid show remarkable concentrations. One such area lies immediately off the northern seaboard of the U.S. This coincides with the position where the warm Gulf Stream meets the frigid Labrador Current, and one suggestion put forward is that large squids are creatures which are sensitive to sudden differences in temperature. If a squid inadvertently passes into colder water, it becomes helpless and cannot avoid rising to the surface where it will soon perish.

One of the famous inhabitants of the middle zone is the living fossil fish, the coelacanth, which appeared 350 million years ago. About 220 million years ago it was the most abundant kind of fish in the oceans. Later it apparently disappeared, but one was recognized in 1938 which had been caught near Madagascar. A number of additional specimens have been found since then. In 1972 one was captured alive for the first time.*

The mesopelagic zone is also characterized by the somewhat mysterious deep scattering layers (the DSLs) of the ocean (see also p. 213). These scattering layers are layers of water which give rise to 'phantom' sound reflections from below the surface regions. They were first detected in 1942 during experiments with under-

* The native fisherman who caught the live 10-kilogram (22-pound) female specimen on 22 March 1972, near the Comoro Islands off the west coast of Madagascar, used tuna bait and a stone sinker on a homemade line. He received a reward of £10,000, equivalent to one hundred years' normal income.

water submarine detection apparatus. During a series of mid-ocean tests the experimenters were surprised to note a series of pulses or pings returning as echoes from about 270 metres to 300 metres (900 feet to 1,000 feet) below. It was immediately realized that their origin was not due to echoes from large objects such as submarines, for these have characteristic echoes which are 'hard' while the anomalous echoes were diffuse and 'soft'. When automatic pen-recorder traces were examined, the echoes were revealed as heavy shadows.

More mysterious still was that the echoes showed diurnal variations. At sunset the reflecting layers appeared to rise nearly to the surface and then suddenly diffuse. At sunrise the layers reformed and immediately began to descend to the normal daylight layers at a rate of about 8 metres (25 feet) per second.

When examined in further detail, it was found that the echoes appeared to originate from three distinct layers, and since then occasionally up to five layers have been noted. The DSLs never appear to cross over but always remain independent, discrete layers. An origin ascribed to sharp differences in ocean temperature was soon ruled out, and speculation was eventually narrowed to their being caused by reflection from marine creatures. But this assumption created further speculations about the physiology of animals which could cause the DSLs, since they appeared to flit from one distinct middle-depth environment of high pressure and low temperature to one of low pressure and high temperature in a cyclic daily rhythm.

During the daylight hours the DSLs are found at depths between 200 and 700 metres (700 and 2,400 feet). At night they either disappear completely or remain as a broad echo-band in the upper 150 metres (500 feet). In different locations the DSLs lie at different depths, but in most places in the ocean they have three layers, the lower one having a depth of about 570–600 metres (1,900–2,000 feet), which actually coincides with the known limits of the twilight zone of marine animals. Occasionally layers can be detected with apparently no diurnal variation – probably representing non-mesopelagic organisms that remain in the twilight zone such as the pelagic worms.

The marine creatures forming the DSLs are probably lantern fish, so named because of their luminous body spots,* and the shrimp-like euphausiids and sergestids. This idea is confirmed by night catches in nets towed near the surface. The lantern fish may be very dominant in producing echoes, for they have swim bladders,† each of which contains a minute bubble of air that may resonate with underwater sound waves to produce detectable scatters.

The deep scattering layers show no differences in depth against variation in latitude. The only difference is that towards higher latitudes species populations change. Significantly, the DSLs are extremely weak in the lifeless desert regions of the Pacific Ocean,‡ and in the sterile wastes of the Arctic Ocean, which are devoid of plankton life, there is no trace at all.

However, like the legendary phantom monsters of the deep, the life of the scattering layers avoids capture on film or by underwater television camera scans. So far the only direct glimpses of these elusive creatures have been through the porthole windows of submersibles during deep dives. Even with the use of submersibles it is difficult to assess the populations at various levels. Evidence so far suggests that the marine organisms comprising the DSLs are not in sufficient concentrations to make worthwhile their economic exploitation as a new sea fishery.

The *bathypelagic zone* and the *benthic zone* represent the environments immediately above and on the seabed, and support an animal population even at the lowest ocean trench deeps 10,800 metres (36,000 feet) below the surface.§ Bottom-feeding animals are

* Lantern fish have long been observed to stay down below during daylight and then rise to the surface at night. They can be observed on the ocean surface by their glow, particularly by their tail lights, which are very bright. One species of lantern fish has lights on its tongue.

† Some fish utilize swim bladders by which they can adjust their buoyancy at different depths. Air is forced into the bladder by secretions from the blood and at ordinary rates of ascent can normally be withdrawn again. If, however, the fish rises to the surface too quickly (if it is caught), the internal air pressure swells the bladder so that it is forced out of the mouth of the fish.

‡ Recently giant amphipods (a variety of crustaceans) more than 28 centimetres (11 inches) long have been found at a depth of 5,000 metres (16,400 feet) in one of the least productive parts of the Pacific.

§ When Jacques Piccard and Lt Donald Welsh of the U.N.S.N. descended to 10,740 metres (35,800 feet) in 1960, they saw in the beam of their searchlight probe a flat fish of the sole family about a foot long and half as wide.

supported by the constant rain of detritus. In the bottom zone, oxygen is quite abundant, and the solid bottom, as a receptor for falling food, helps the highly adapted population to eke out a living (see pp. 203–4).

Marine life adaptation at the bottom of the ocean has attained a very high level. Some fish attract prey with luminescent lures or by the practical art of mimicry. Species have greatly extensible jaws, and expandable abdomens are a common feature. But by surface and middle layer standards, the fishes of the bottom zones are Lilliputian monsters not more than half a foot in length on average. In the great depths *bioluminescence* has reached an advanced level of evolutionary development, and more than two-thirds of the species are able to produce light.

Some of the fishes, euphausiids and squids are equipped with searchlights complete with reflector and lens. Some fish are also able to discharge luminous clouds to blanket their escape. Visual organs are arranged on some creatures so that they can be used to counter-shade their silhouettes against faint light sources from above. Lights are also used to attract mates and act as food baits. The angler fish are as wide as they are high. They have received their name from their incredible methods of catching prey: they move slowly through the water, and use a trailing rod and line with a worm-like bait on the end of it (Figs 35 and 63). Another deep-water fish is the sabre-toothed viper fish which has luminous patches in its mouth that act like lures to smaller creatures.

Typical of bottom fish is the gulper, or pelican eel, found below 2,000 metres (6,500 feet) in tropical or subtropical waters. This fish is all head (Fig. 35); although it is 25 centimetres (10 inches) long, its brain case measures only 6 millimetres ($\frac{1}{4}$ inch).

Another 'fossil' deep-sea creature is the primitive mollusc *Neophilina galathea*, which is neither clam nor snail and was thought to have become extinct 350 million years ago (at the end of the Devonian period). In 1952 one was caught in 3,600 metres (12,000 feet) of water during the cruise of the research ship *Galathea* while it was dredging in deep water off the Pacific coast of Mexico.

About 2,000 species of fish populate the bottom zones, and genera such as the bristle-jawed fishes may in fact be the most numerous fishes in the world. However, in relation to the volume of water it represents, the bathypelagic zone animal population is very sparse, and its calculated *biomass* has been estimated at only one-hundred-millionth of its water volume.

Creatures that live permanently on the bottom are living representatives of some of the earliest multi-celled creatures to exist on Earth. We find glass sponges, sea lilies (crinoids) and lamp shells (brachiopods). Some of the juveniles of the abyssal population actually develop at intermediate levels of the ocean and then later return to the bottom (see pp. 211–13).

Bottom-living creatures are subject to pressures sometimes greater than 1·1 tons per square centimetre (7 tons per square inch). How such creatures can survive used to be a mystery to marine biologists. Nowadays, however, it is appreciated that most of the substance of a bottom creature is pervaded with water; water is practically incompressible, and its cushioning properties protect the living creature from harm.

Environmental Sounds

The environment below the waves is not the totally silent zone of man's own immediate experience when he submerges below the surface. Hydrophones placed below the surface pick up many different sound patterns, some of which undoubtedly represent communication between marine creatures. Commonly heard are ghostly moans, whistles, squeals, boops, honks, cackles, moans, shrieks and groans – providing a wide-ranging vocabulary of sound noises.

Most of these sounds are probably associated with hunting and mating, but the whales as a specialized marine family appear to have developed sound communication to a high level of sophistication. Dolphins (a variety of toothed whale) have been subject to intensive field-study, and they seem to use a form of sonar to probe the shape of objects. Signals from dolphins have been measured very accurately which show a range of wave-lengths from 200 Hz to 150,000 Hz. It has been suggested that dolphins possess several transmitters and receivers which allow them to acquire acoustic echoes from an object at several frequencies. As they are animals that are continually on the move, the purpose of the multi-signals is probably to provide changing angular information about an object. A three-dimensional object would scatter acoustic transmissions in different ways. The dolphin's acoustic receiver might then reform them with a central computer-like ability and flash a composite picture to the dolphin's brain.

Large whales of all species also appear to be capable of a sophisticated repertoire of sounds. Nineteenth-century whalers were frequently witness to whale cries, when the great beasts lay in their death throes alongside. Modern study of whales with hydrophones has revealed that whales may actually sing in the literal sense. A gramophone record has been on sale commercially for a number of years entitled *Songs of the Humpback Whale* which contains some of the most fascinating and beautiful sounds of the nature world ever recorded. The sounds are eery and electronic-like, and somewhat reminiscent of the cosmic sounds which emerge from the amplifiers of radio telescopes. It is likely that whale 'songs' perform a precise social function. Their most important functions are probably to help maintain contact within the school and to relay information about predators and plankton food supplies.

P.L.B.

7 ISLANDS AND CORAL REEFS

Islands have long held a fascination for mankind, and in world folklore legend they figure prominently. Plato described a great empire island continent called Atlantis whose inhabitants, in about 9000 B.C., were supposed to have conquered all the known world, except that of the Greeks. But Atlantis was said to have been engulfed overnight, and it disappeared without a trace.

The legend of the supposed lost Atlantis has continued up to the present time, and rarely a week goes by without some reference to it or some new theory being evolved to explain its disappearance, or its presumed location on the seabed.* There may be some truth in the Atlantis legend, leastwise that part of it concerned with its sudden disappearance. Islands certainly do disappear – practically overnight – particularly volcanic islands whose unconsolidated ash bases have

* Some legends give its location in the Mediterranean (the island of Thira? – see below), others place it in the North Atlantic.

insufficient structural support to resist the onslaught of a tsunami caused by an earthquake or volcanic eruption * (see p. 45).

The islands of the seas and oceans fall into two physical classes, the continental and the oceanic. Continental islands, as the name implies, are situated close to the continental masses and geologically represent an uplifted part of the continental shelf. They are made up of rocks similar to those found on the adjacent continent. The British Isles and Newfoundland are good examples of continental islands. Oceanic islands usually rise abruptly from the deep water of the ocean, and they are closely related to submarine ridges such as the ridges giving rise to the islands like the Azores, Ascension, St Helena and Tristan da Cunha.† Even though some of these islands now appear to be far removed from the present-day Mid-Atlantic Ridge, they were all 'born' on the ridge (see 'Evolution of the Oceans and Continental Drift', p. 48). The oceanic islands are thus volcanic in origin. Coral islands, although sedimentary structures in their upper parts, all have volcanic roots (see below).

From the study of sea-floor spread we see that the islands and continents are continually on the move, being pushed outwards by the inexorable conveyor-belt motions of the emerging ridge material. It has already been noted that owing to this movement, the birth of islands may be approximately dated (p. 58). The Pacific Ocean contains the largest number of islands, and it has been estimated that it contains more than 20,000 – ignoring the smaller islets measuring only a few square metres in area. Nevertheless, islands as a whole represent a very small percentage of the Earth's surface area.

Coral Reefs

When Charles Darwin went aboard the *Beagle* in 1835, little was known about the origin and history of coral islands and reefs. Many of the great navigators have experienced the hazards and dangers of

* One such catastrophe occurred on the island of Thira (Santorini), one of the Cyclades Islands, between Greece and Crete. In about 1520 B.C. a violent eruption snuffed out the brilliant maritime Minoan civilization and brought about a shift of power to the Achaens, the mainland forebears of the Classical Greeks.

† Note that some 'oceanic' islands such as Madagascar and the North and South Islands of New Zealand are continental islands. They are continental fragments separated from their original mass through continental drift plate-tectonic movements.

the partly submerged shoals of coral. In more than one instance the course of maritime history was influenced by the hidden presence of a coral reef – perhaps even more than any other single feature of the oceans. The old navigators learnt to dread the enigmatic ring-shaped masses that inexplicably rose sheer, without warning, from the blue unfathomable depths.

Darwin knew that the peculiar corals could grow only in shallow water, and their presence in mid-ocean could only be explained if it were assumed that they grew on top of huge volcanic peaks which lay hidden below the surface. When these peaks were sufficiently eroded – but not too much – coral would begin to grow.

During the course of the *Beagle*'s voyage, Darwin noted – as Humboldt before him – that the Andes Mountains appeared to be uplifted geological features that had once been part of the seabed. He reasoned that perhaps uplift was balanced elsewhere by a downwarping action, and the ocean basins seemed the ideal place where this might take place. When the *Beagle* reached Tahiti, Darwin's ideas had gelled into a definite theory, and the Subsidence Theory of coral reef formation was born.

At Tahiti, Darwin noted that the volcanic islands were surrounded by an enclosing reef. If the volcanic island were removed from its centre, then all that would remain would be an atoll, and the mechanism to explain the gradual removal of a volcano would be a slow but progressive subsidence with the coral up-growth maintaining a rate to keep pace with the rate of subsidence.

Darwin's sequence of events can be seen in actual examples of volcanic islands scattered about the oceans, but particularly in the Pacific (see Figs 15 and 36). Darwin's theory was later widened to include fringing barrier reefs such as the Great Barrier Reef of Australia.

Darwin's theories of coral reef formation met with some opposition, since some believed that certain reefs of the Pacific could not be explained in this way. An alternative theory proposed, known as the Glacial Theory, sought to explain coral reef formation in terms of cyclic fluctuations of sea-level during the recurrent ice ages. This theory considered that during the ice ages the oceans were so chilled that coral was unable to grow. Coral islands are characterized by the presence of tropical or subtropical water. In recent years evidence has come to light that during the Pleistocene, the ocean

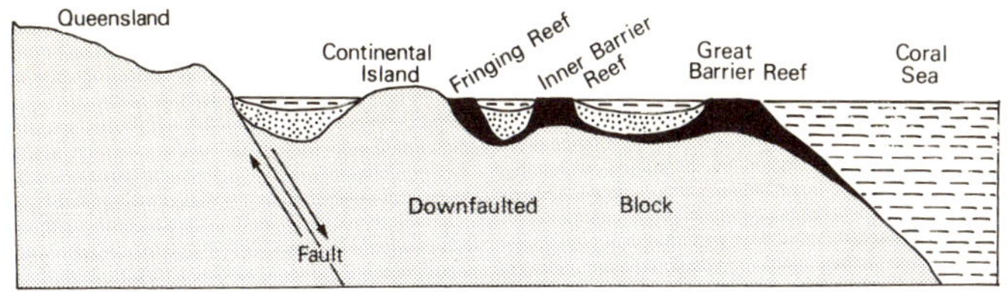

Fig. 14 Schematic section through the Great Barrier Reef (Queensland, Australia) showing the relationship of the offshore reef to the mainland. The coral-reef rock is shown in black; lagoon and channel sediments are shown dotted.

By contrast, most barrier reefs are island-encircling structures forming regular or irregular rings of variable width, usually interrupted by open passages on the leeward side (see also Fig. 36).

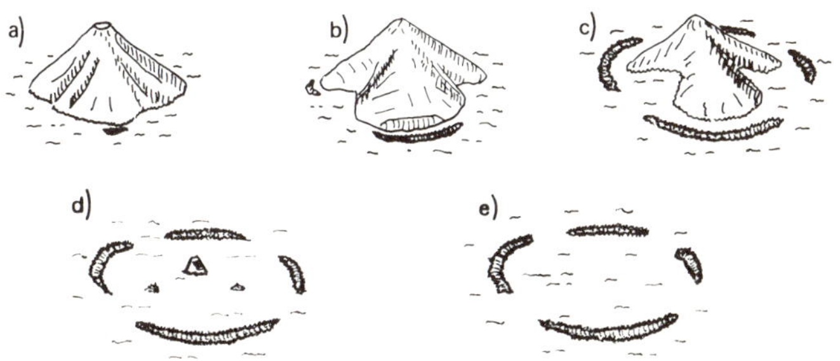

Fig. 15 Examples of various coral islands:
a. Tahiti; Grand Comoro; Hawaii. b. Rarotonga; Oahu. c. Santa Cruz; Mayotte. d. Truk; Aitutaki; Clipperton. e. Bikini; Eniwetok; Kwajalein.

temperature fell by as little as 4–7° C. (40–44° F.); although this would reduce the total global area where corals might thrive, it would still allow reef growth in a restricted belt.

Murray, from his own observations while aboard the *Challenger*, also put forward an alternative theory to Darwin's which did not invoke subsidence. According to Murray, the base of a coral reef consisted of a submarine hill rising from the ocean bed – probably a denuded sub-surface volcanic peak – which was then built up from pelagic deposits until it reached a depth of about 55 metres (180 feet) below the ocean level, at which point a coral reef would then begin to grow.

Proof that Darwin's theory was the correct one did not come until 1952, when deep borings made at Eniwetok Atoll in the Pacific indicated a thickness of over 1,350 metres (4,500 feet) of coral rock before the base rock of volcanic basalt was reached – thus indicating a build-up over a long, slow subsidence of its foundation.

Present-day corals that build reefs live in warm water and rarely grow in waters cooler than 20° C. (68° F.), but exceptions do occur.* Most coral reefs are situated between latitude 30° North and 30° South of the Equator, but again there are exceptions such as the Islands of Bermuda (32° North) due in this case to the warming influence of the Gulf Stream.

In the Southern Hemisphere extensive coral growth favours the eastern seaboards of the continents. In Australia the Great Barrier Reef lies along the eastern seaboard, but corals also grow on the western seaboard, although here they are much less extensive.

Live corals do not like to be exposed and are not usually found above present-day low tide levels. Neither do they grow in water depths exceeding 45–55 metres (150–180 feet).† They also require unclouded, well-oxygenated water and quickly die if exposed to fresh or silt-laden waters. Since food supplies are generally more plentiful on the seaward side of a coral reef, the corals tend to grow outwards. But a study of the total evolutionary changes observed in growing reefs must take into account wave-action, drifting sand and the accumulation of vegetation.

The *fringing reef* consists of an even platform of coral, fringing a coastline with a shallow, narrow lagoon between it and the mainland, with its seaward edge sloping down towards the deep water. The *barrier reef* is separated from the mainland by a much deeper and wider channel such as is exemplified by the Great Australian Barrier Reef off the coast of Queensland.

The coral *atoll* consists of a coral reef which may be circular, elliptical or horseshoe-shaped, enclosing a lagoon without any central island.

Corals consist of many species. Coral rock is formed by small marine animals called coral polyps which grow in clusters in as-

* As for example on the Abrolhos Islands off the coast of Western Australia.

† Not all corals are *reef* builders, and some live at great depths; many of them in total darkness such as *Flabellum* (found 3,300 metres (10,800 feet) down).

sociation with algae. The polyps, with the assistance of the algae, produce hard skeletons of calcium carbonate, and as each coral dies, it adds to the accumulation to form a mass of rock known as coral limestone.

The coral polyp is shaped like a double-walled cylinder tube which is closed off at one end and open at the other where a wreath of tentacles surrounds the mouth. The most significant characteristic is its ability to secrete lime, which forms a hard cup about itself in much the same way that a mollusc secretes its hard shell. Most corals build colonies of many individuals, but all individuals are derived from a single fertilized ovum that matures and then forms new polyps by budding. Corals take on different colours from the minute algae plant cells that reside in the soft tissues in a relationship of mutual benefit. They supply the coral with oxygen and remove unwanted phosphate, ammonia and carbon dioxide.

Corals can grow only to the level of low-water neap tides. Coral branches generally grow at the rate of about 1 centimetre ($\frac{3}{8}$ inch) per year, but tips of lighter branching corals may grow up to 10 centimetres (4 inches) per year.

Many of the coral reefs of the oceans are quite old. Radiometric dating indicates that the Pacific, Indian and Caribbean reefs have ages ranging from 40,000 to 160,000 years for coral reefs at present above sea-level. Reefs, however, are quickly broken down. They may be bored by sponges, numerous varieties of worm, molluscs and algae, and extensively fragmented by wave-action. The Great Barrier Reef in Australia and other great reefs of the world have suffered greatly in recent years from the attack of the Crown of Thorns starfish (*Acanthaster planci*). In Australian waters, the starfish was seldom seen until the late 1950s when it began to appear in increasing numbers round Green Island near Cairns. The creature measures 46 centimetres (18 inches) across and has up to sixteen thorn-covered arms. It attaches itself to the living tips of coral and consumes the coral's tissues through tiny suckers. In the 1960s it caused massive destruction of reef coral, reducing a beautiful coral seabed to masses of underwater rubble. In Australia, the main predator of the starfish is the triton, or trumpet shell. These were formerly collected in large numbers by tourists visiting the reef, which may have allowed the Crown of Thorns to multiply. Restricting the taking of tritons appears at present to be the only effective measure against the potential threat of the destruction of the entire 1,900 kilometre (1,200 mile) long coral mass, forming the most beautiful reef in the world.*

Corals which grow in the hurricane belts suffer severe wave erosion, and, once exposed, the surface of the reef may be completely swept away. Although reefs may recover in a period of a decade or two after such a catastrophe, those which lie within well-defined, present-day storm tracks cannot because of the frequency of storms. Nevertheless, the coral debris arising from coral reef erosion often gives rise to islands and cays which 'grow' further due to the accumulation of vegetable matter, bird lime and other biological debris. In addition tropical sea-water is saturated with calcium carbonate, which is precipated in the intertidal zone round such islands and assists island building by forming a solid beach rock. Islands formed in this manner show an extremely complicated geological history.

Corals have been extensively used in geophysical studies. The fine ring-like banding which is displayed by the skeletons of some corals show variations in growth which have been used to determine the Earth's rate of rotation in past eras. Corals show extremely delicate responses which are reflected in their ring-like bandings. By comparing present-day corals with fossil corals, variations in band growth can be directly related to the change in the length of day. It is an established fact that the length of the day has been changing – by something like two seconds every hundred thousand years – as a result of the tidal friction produced in the oceans by the Moon. If the rate was constant, this would imply that in the Cambrian period (570 million years ago) there were about 428 days in each year. Fossil corals of the Middle Devonian period reflect values between 385 and 405 days, which is close to the computed time of about 400 days for the Middle Devonian period (370 million years ago).

Not all corals are reef builders. One group of corals, *Flabellum curvatum*, lives in deep water and has been recorded at depths of

* Recent studies indicate that it may be possible to control the Crown of Thorns starfish by feeding them contraceptive agents! In addition it is now known that the starfish has two other natural predators: the lump-headed wrasse (*Cheilinus undulatus*) and the trigger fishes.

3,300 metres (11,000 feet). These corals do not have a symbiotic relationship with algae, and in spite of their dark habitat show vigorous growth bands.

One reef-building coral that grows in north-west Australia shows peculiar magnetic qualities and is highly influential on ships' compasses.

* * *

Island environments are extremely important in the study of the processes of evolution. Charles Darwin realized this during his visits to the Galapagos Islands while aboard the *Beagle.* Although many of the islands had animals which were obviously derived from a common evolutionary stock, they had diverged considerably after becoming isolated. Island habitats throughout the world show similar divergencies. Animal and plant life often develops in peculiar and often bizarre forms when evolved in isolation, and unique species occur such as the dodo on Mauritius, the moas in New Zealand, the elephant-bird in Madagascar and the coco-de-mer palm in the Seychelles. In the Monte Bello Islands off north-west Australia (once joined by a land bridge to the Australian continent) wallabies have developed a claw in their tails, which from an evolutionary point of view has been interpreted as providing an environmental advantage in a habitat which radically changed.

In the present-day world, islands, because of their isolation and simplicity, make ideal experimental laboratories for studying the principles governing ecology. In the past most island experiments were the results of 'experiments' arranged by Nature herself. The future trend in ecological studies is towards extending the laboratory analogy directly to islands. One such example already operational is that on Aldabra Atoll in the Indian Ocean where a research station is maintained by the Royal Society.

P.L.B.

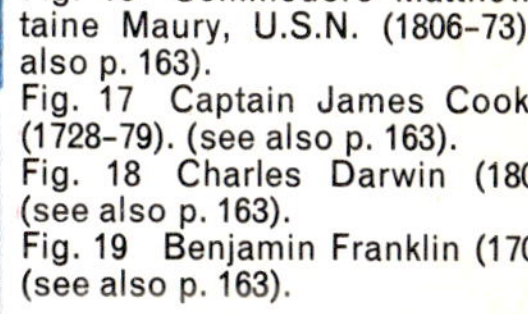

Fig. 16 Commodore Matthew Fontaine Maury, U.S.N. (1806–73). (see also p. 163).

Fig. 17 Captain James Cook, R.N. (1728–79). (see also p. 163).

Fig. 18 Charles Darwin (1809–82). (see also p. 163).

Fig. 19 Benjamin Franklin (1706–90). (see also p. 163).

Fig. 20 Deep-sea deposits (sediments) (see also pp. 64–6).

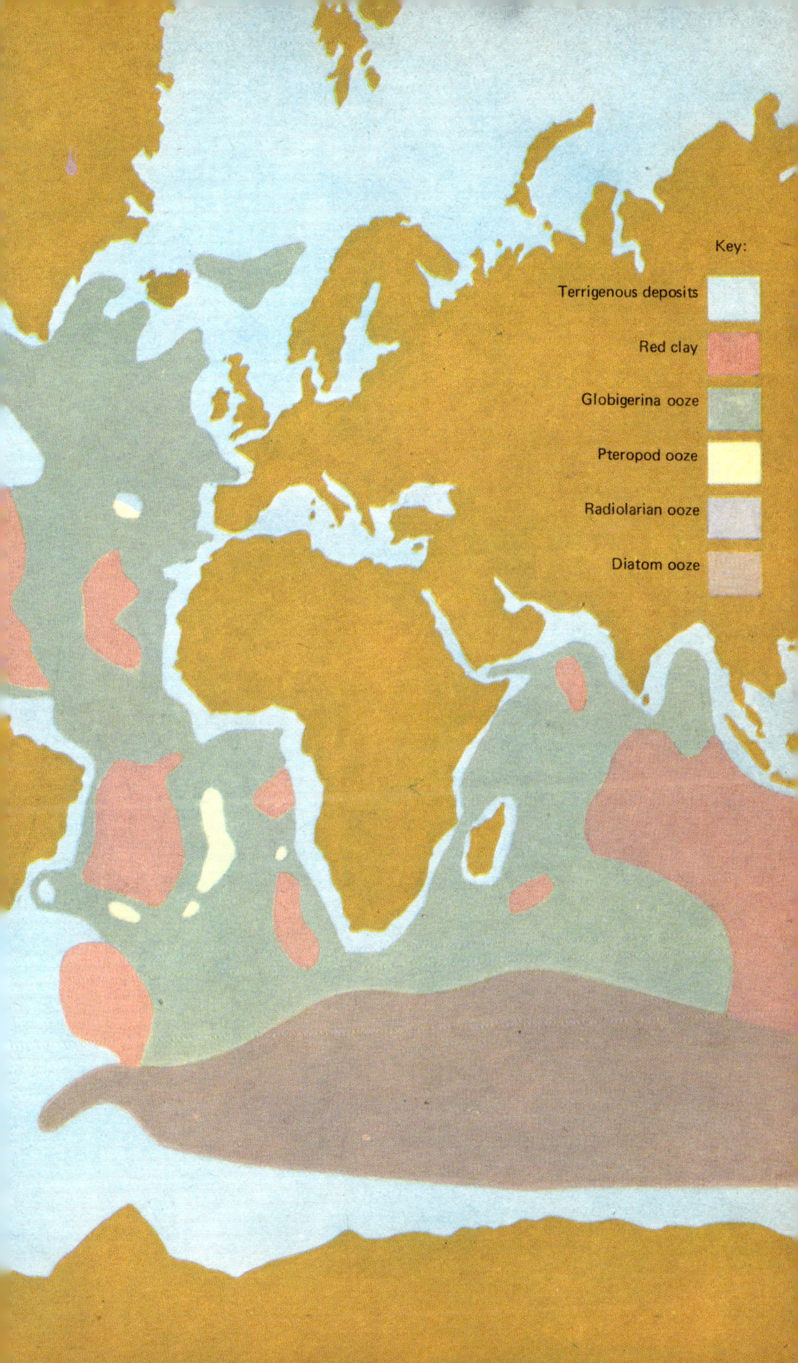
Key:
Terrigenous deposits
Red clay
Globigerina ooze
Pteropod ooze
Radiolarian ooze
Diatom ooze

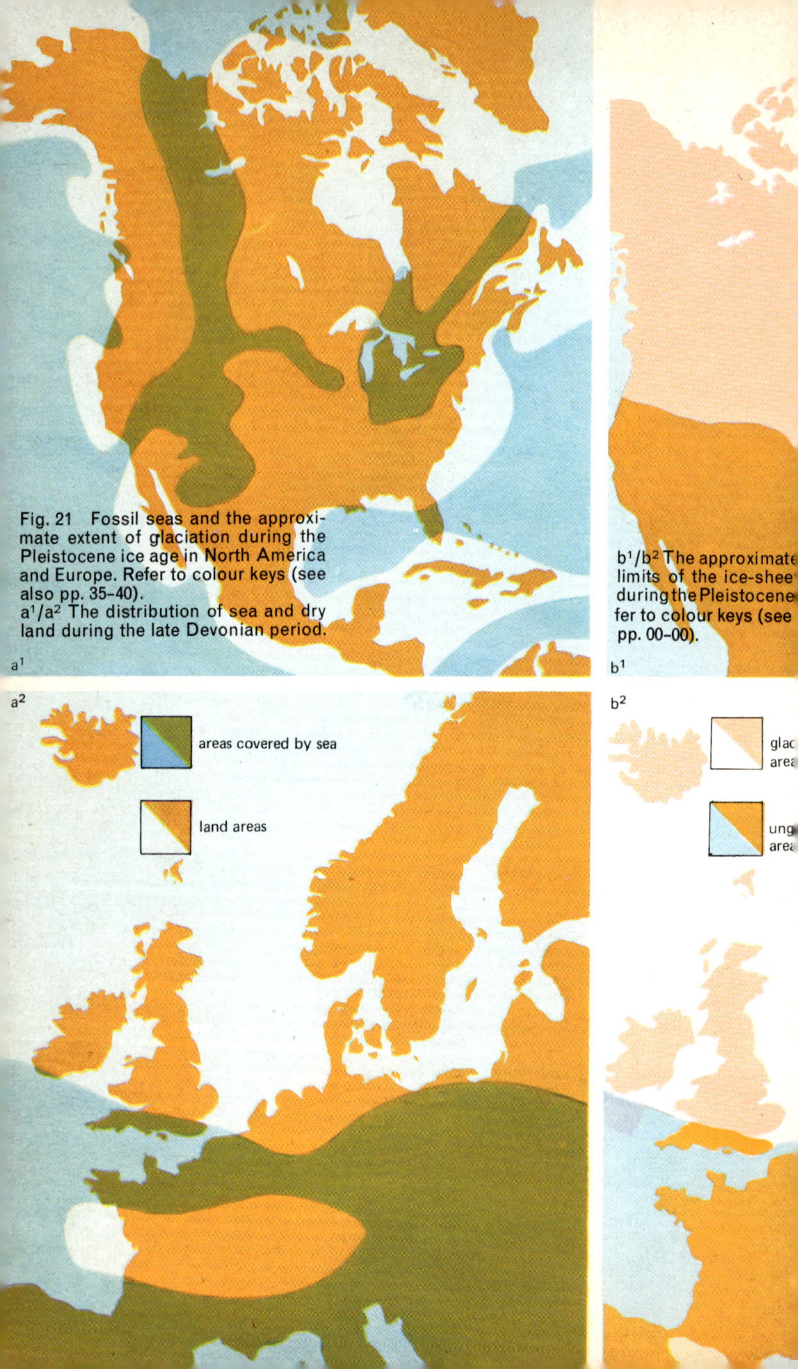

Fig. 21 Fossil seas and the approximate extent of glaciation during the Pleistocene ice age in North America and Europe. Refer to colour keys (see also pp. 35–40).
a[1]/a[2] The distribution of sea and dry land during the late Devonian period.

b[1]/b[2] The approximate limits of the ice-shee during the Pleistocene fer to colour keys (see pp. 00–00).

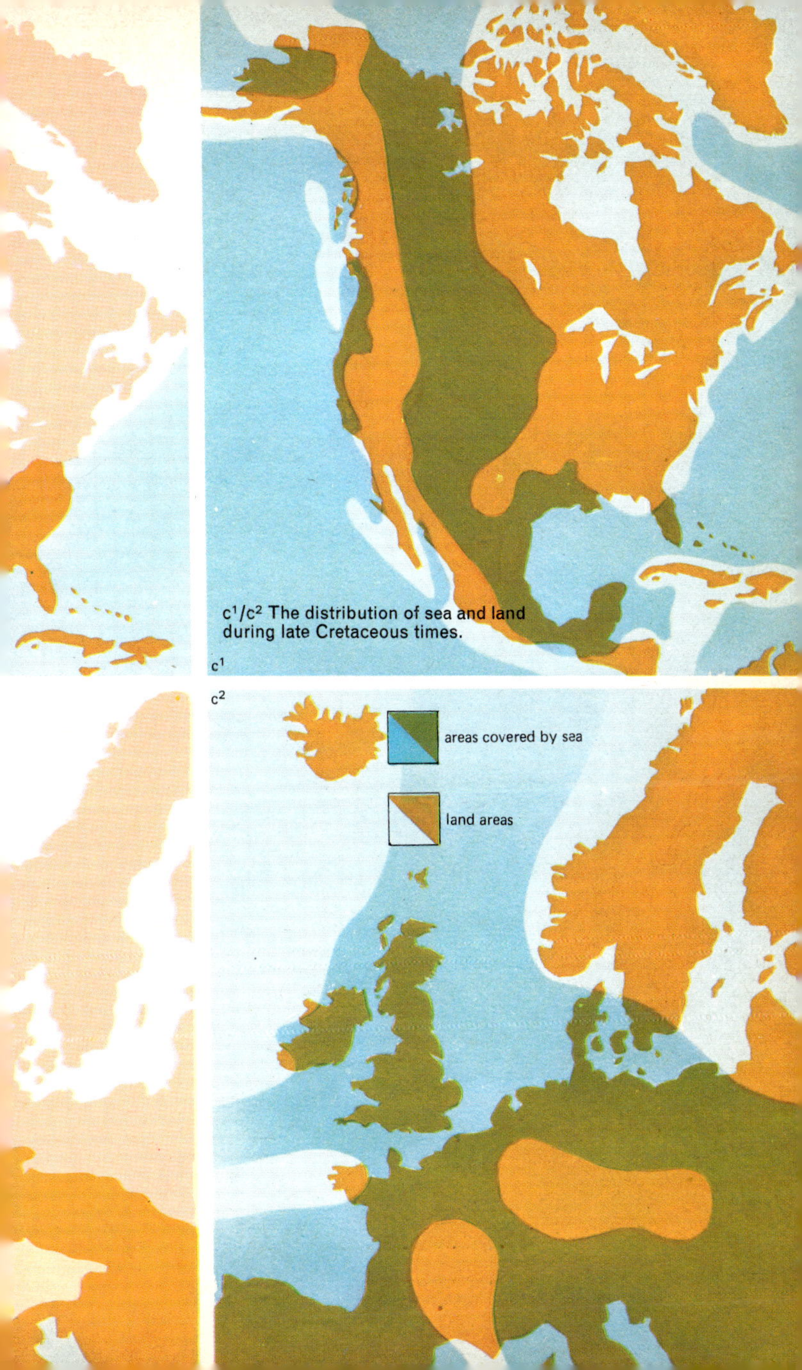

c^1/c^2 The distribution of sea and land during late Cretaceous times.

Fig. 22 The distribution of continental shelf area and deep trenches (deeps). In several areas, at the termination of the shelf, there are steep cliff-like features where the water deepens rapidly (see also Figs 23 and 30 and pp. 12–14).

Fig. 23 The Blake Plateau, an area 800 metres (2,650 feet) deep off the coast of Florida, which is strewn with iron-rich nodules. The lighter-coloured area of the seabed adjacent to the present-day coastline represents the approximate limits of the 200-metre continental shelf whose sea edge was the North American coastline c. 15,000 years ago (see also pp. 12–14).

ARCTIC OCEAN
ASIA
Oyashio
Alaska Current
NORTH AMERICA
North Pacific Current
California Current
Kuroshio
NORTH PACIFIC OCEAN
North Equatorial Current
Equatorial
Counter
Current
South Equatorial Current
Peru
AUSTRALIA
E. Australia Current
SOUTH
PACIFIC
OCEAN
West Wind Drift
ANTARCTIC OCEAN

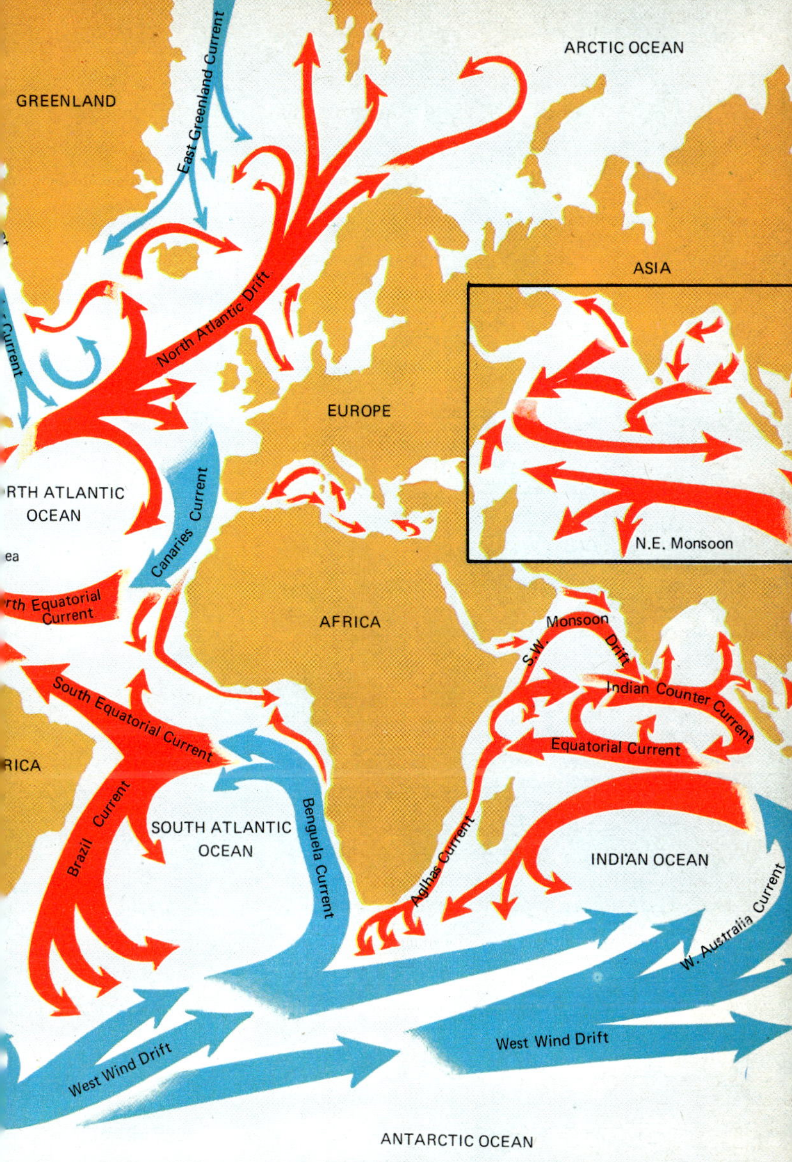

Fig. 24 The world's surface currents. Warm currents are shown in red, and cold currents in blue. Note that the Peru Current is nowadays the more usual name for the Humboldt Current. Inset (*above right*) shows how some currents in the northern Indian Ocean change direction during the period of the North East Monsoon (see also Figs. 25–6 and pp. 26–31).

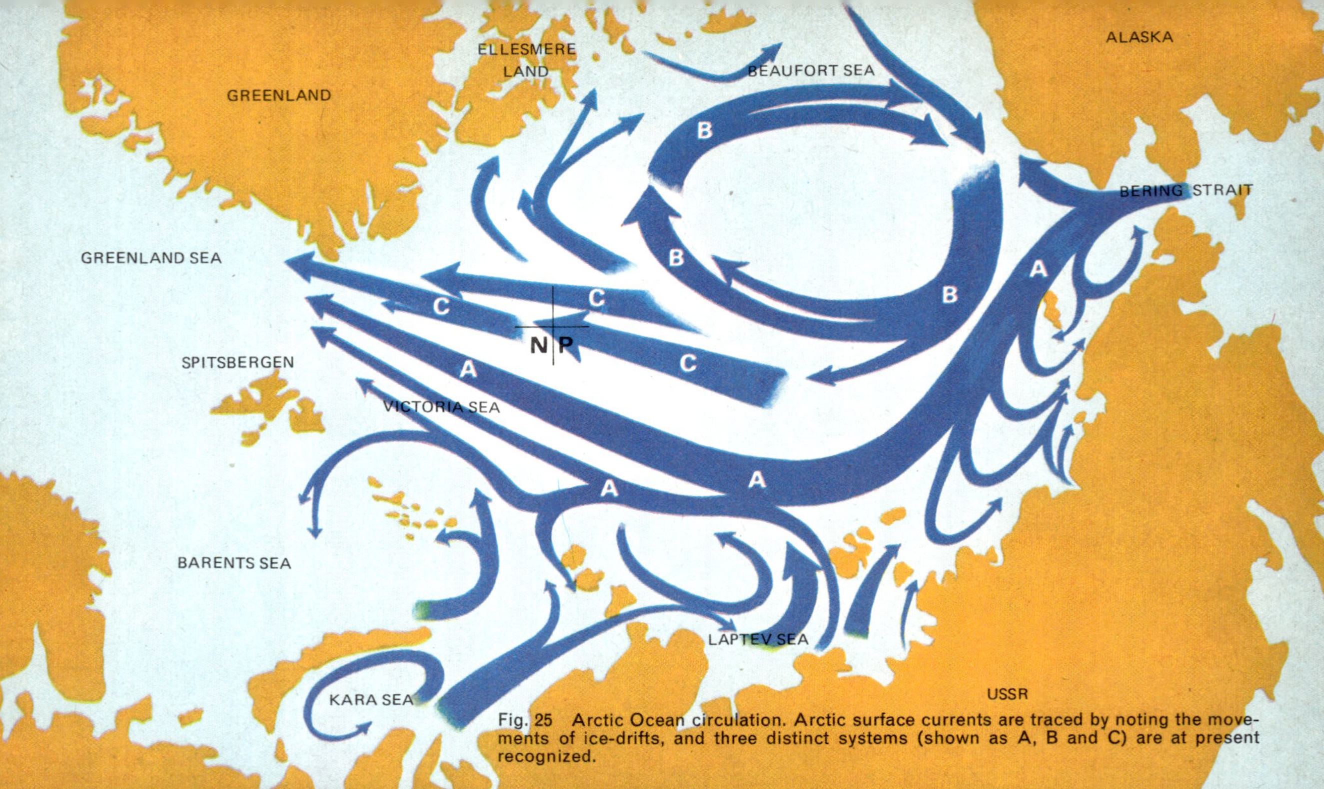

Fig. 25 Arctic Ocean circulation. Arctic surface currents are traced by noting the movements of ice-drifts, and three distinct systems (shown as A, B and C) are at present recognized.

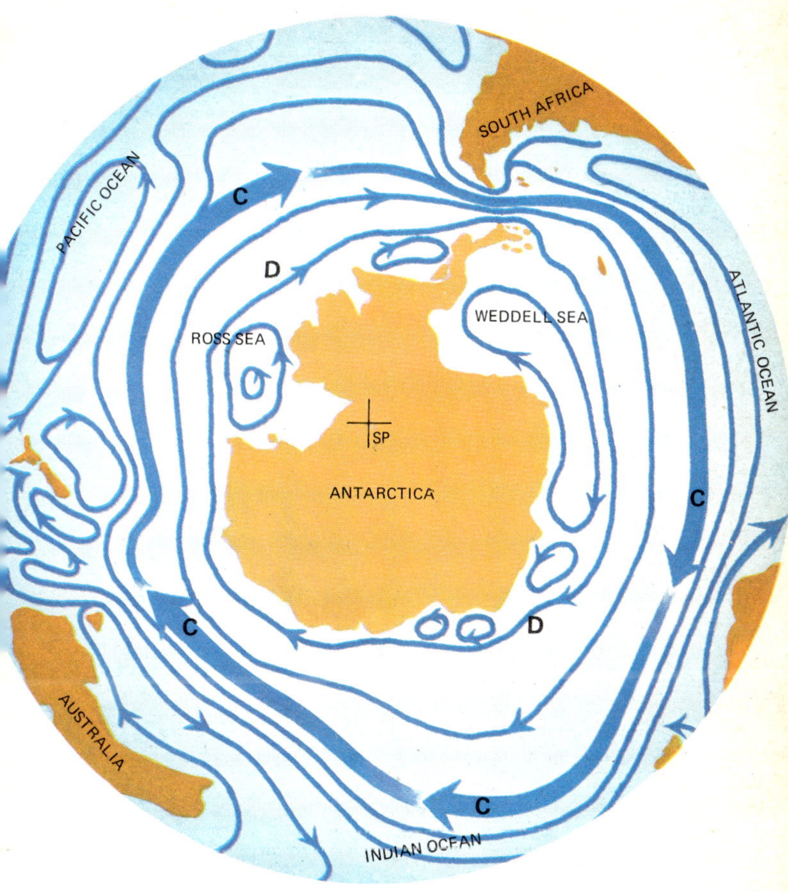

Fig. 26 Antarctic circulation. The main Antarctic current (the West Wind Drift) flows without interruption in a clockwise direction. The heaviest flow of water is in the region of the Antarctic Convergence (C) where both temperatures and salinity change abruptly. The Antarctic Divergence (D), near the coastline of Antarctica, is the region of minimum flow (see also pp. 21–2 and 30).

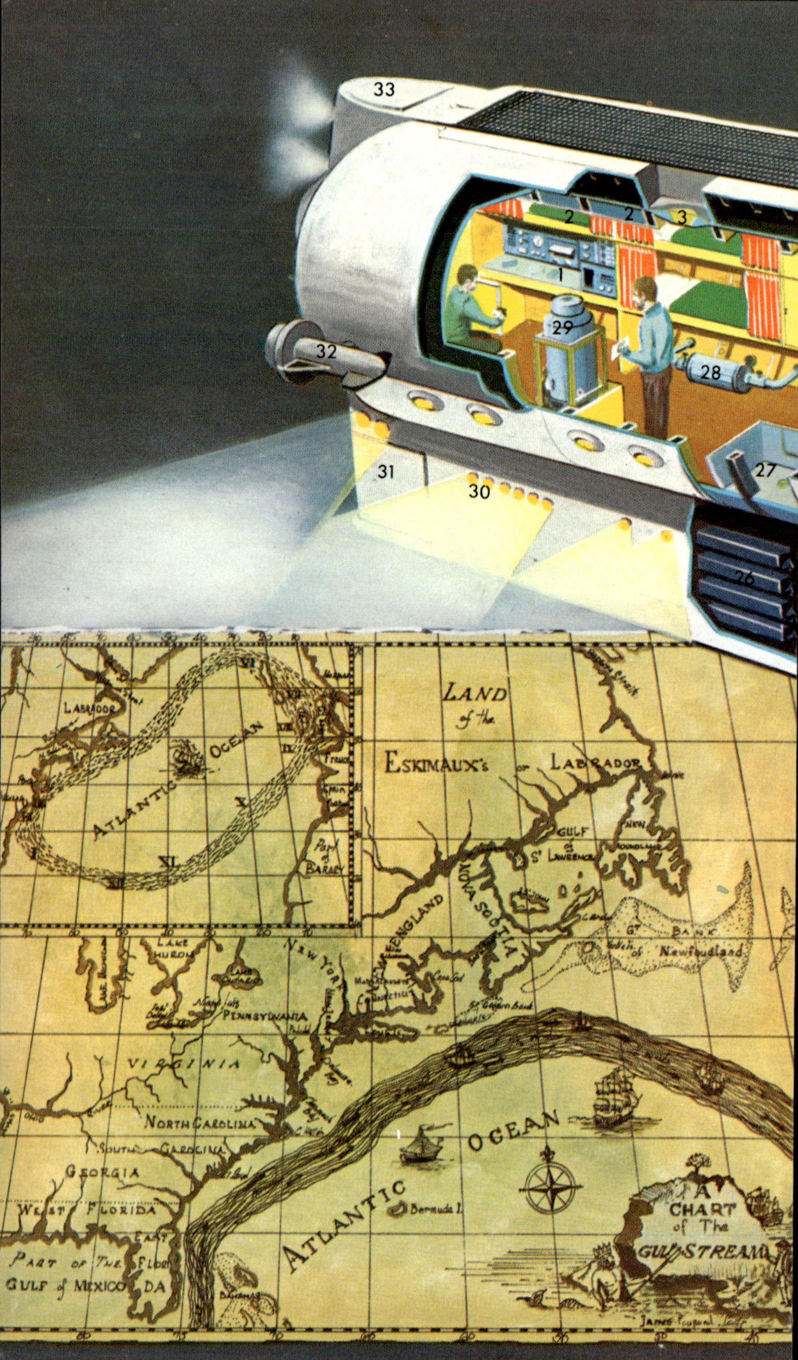
33
2
2
3
1
29
32
28
31
30
27
26
LABRADOR
ATLANTIC OCEAN
LAND of the ESKIMAUX's or LABRADOR
GULF of St LAWRENCE
NOVA SCOTIA
NEW ENGLAND
Gt BANK of Newfoudland
LAKE HURON
NEW YORK
PENNSYLVANIA
VIRGINIA
NORTH CAROLINA
SOUTH CAROLINA
GEORGIA
WEST FLORIDA
EAST FLORIDA
PART OF THE GULF of MEXICO
OCEAN
ATLANTIC
Bermuda I.
A CHART of The GULF STREAM

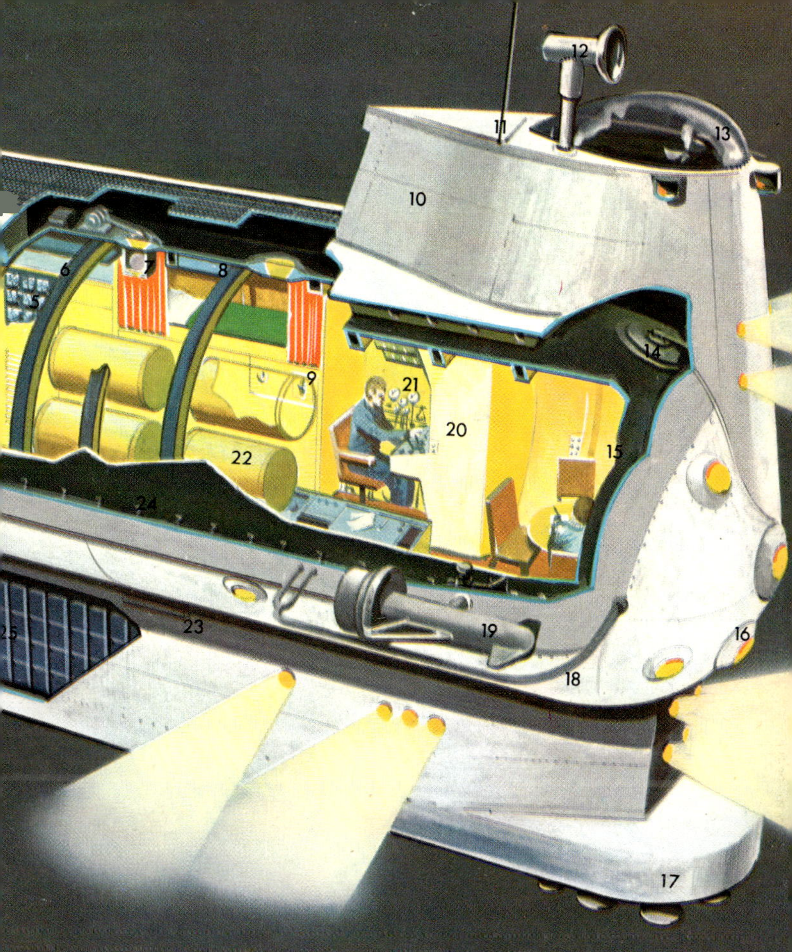

Fig. 27 (*above*) The 130-ton submersible *Ben Franklin* designed by Jacques Piccard and built by the Grumman Aircraft Engineering Corporation. In 1968 it successfully completed a 33-day, 2,400-kilometre (1,500-mile) submerged drift experiment along the Gulf Stream from West Palm Beach, Florida, to Cape Hatteras, North Carolina (see also p. 31). (*left*) Benjamin Franklin's chart of the Gulf Stream, published in 1770, to help mariners avoid the coastward-flowing current and thus save two weeks' sailing westwards (see also pp. 30–1).

1 Scientific experiments control centre. 2 Cold-water tanks. 3 Observation port. 4 Iron shot ballast. 5 Innerters. 6 CO_2 scrubber. 7 Canister release air lock. 8 Cold-water tank. 9 Electrical switch panel. 10 Conning tower. 11 Surface radio antenna. 12 TV cameras. 13 Surface lookout. 14 Hatch. 15 Wardroom. 16 Observation port. 17 Instrument package (cameras, strobe lights and hydrophones). 18 Motor guard. 19 Propulsion motor. 20 Control console. 21 Hydraulic and pneumatic controls. 22 Super-insulated hot-water tanks. 23 Sonar. 24 Ballast tank. 25 Batteries. 26 Battery reservoirs. 27 Shower. 28 Transparent quartz tube. 29 Oxygen tank. 30 Lights. 31 Rudder. 32 Propulsion motor. 33 Access to aft hatch.

Fig. 28 Deep circulation in the oceans showing the world-wide flow pattern modelled after the ideas of Henry Stommel. The darker blue areas represent the deeper parts of the ocean basins.

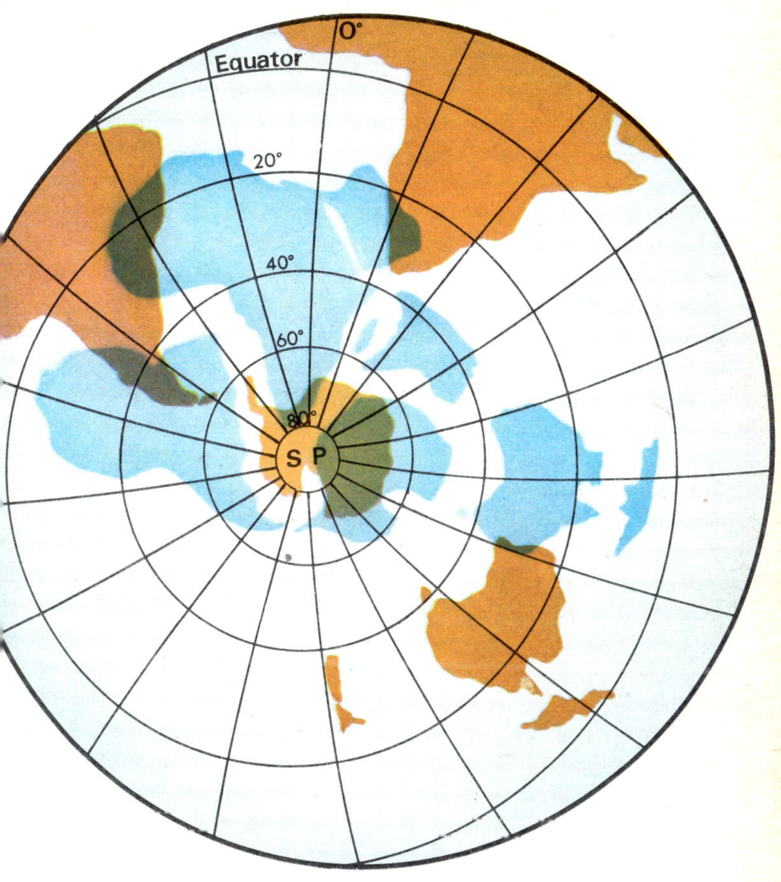

Fig. 29 Continental drift. A reconstruction of Gondwanaland during earlier geological times showing the location of the various continental land masses depicted in light blue; present-day positions are depicted in brown (see pp. 48–64).

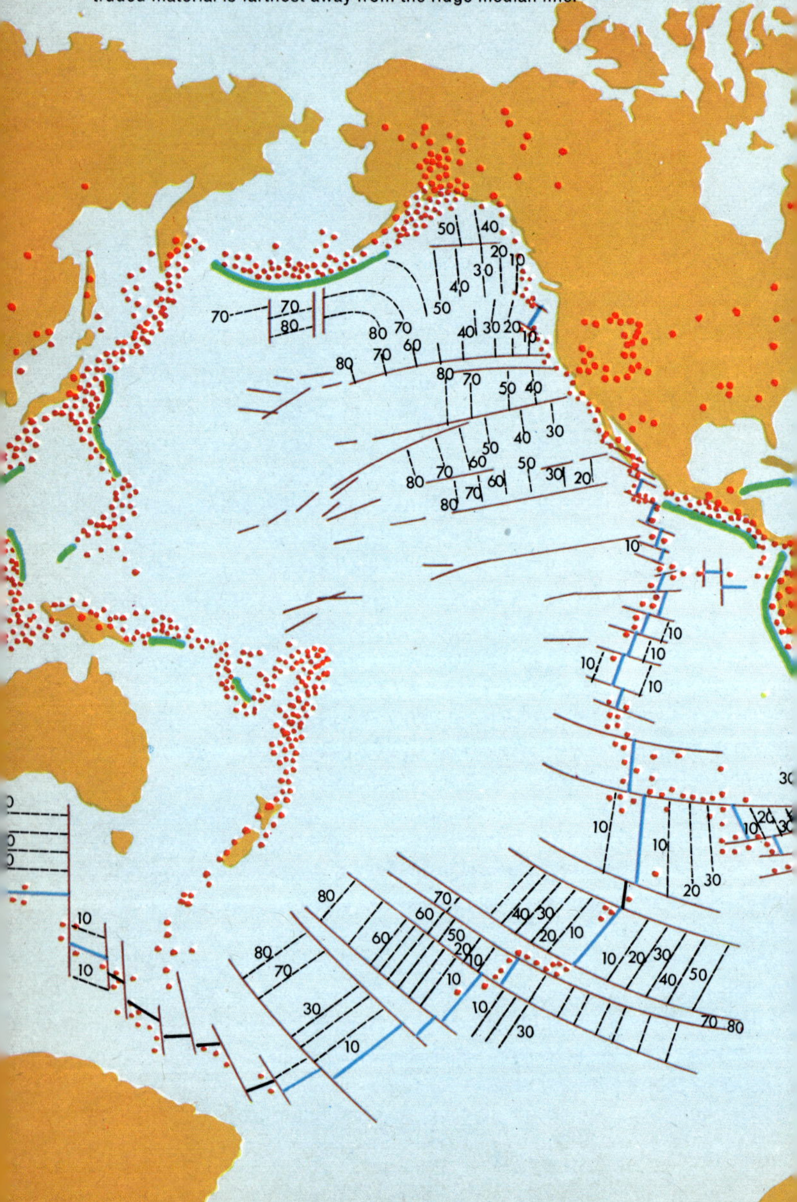

Fig. 30 The mid-ocean ridge system (blue lines) and active earthquake epicentres (red dots). The numbered pecked lines, depicted parallel with the main ridge system, represent the present position of material extruded from the ridge and are dated in unit intervals of 10 million years. These dates clearly indicate that the oldest extruded material is farthest away from the ridge median line.

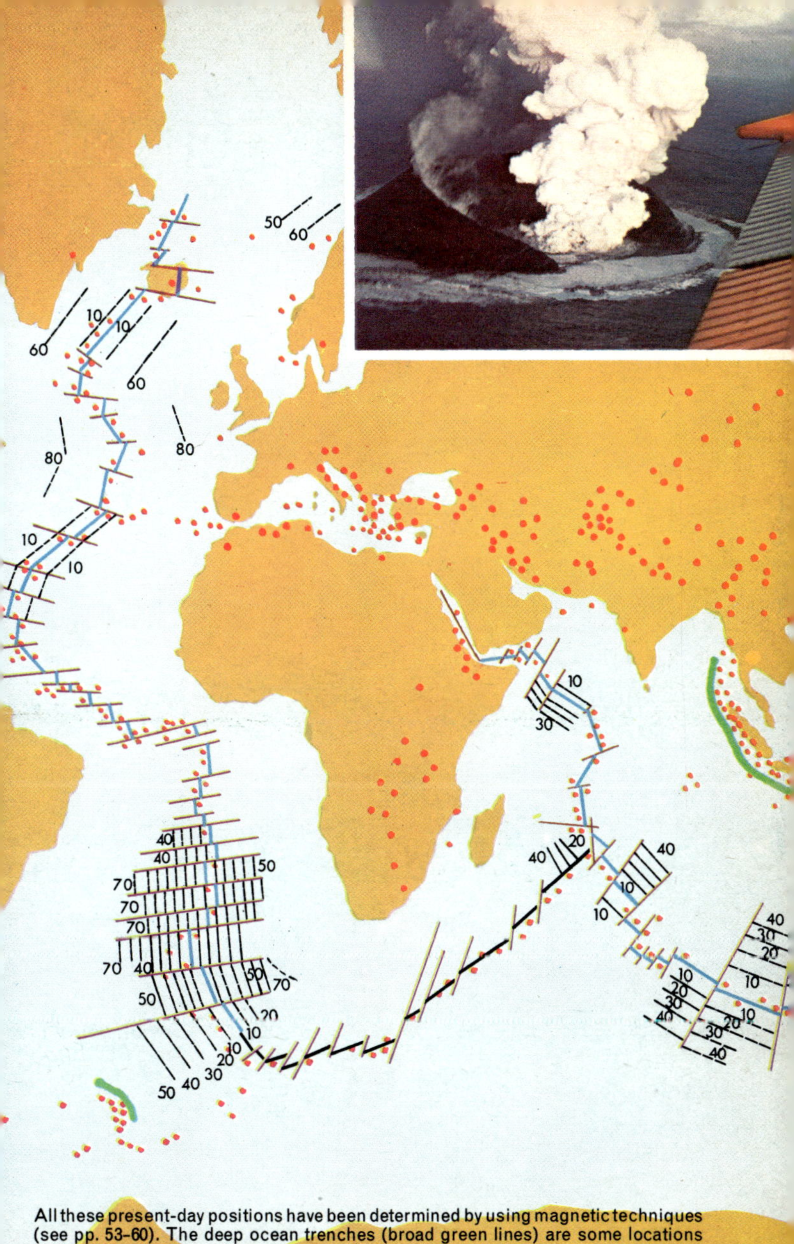

All these present-day positions have been determined by using magnetic techniques (see pp. 53–60). The deep ocean trenches (broad green lines) are some locations where the spreading ocean floor descends back into the Earth's crust (see also Fig. 10 and pp. 48–64). The dark red lines represent the boundaries of the principal fracture zones in the crustal rocks. (*top right*) Sea-floor spread created the new volcanic island of Surtsey off the coast of Iceland in 1963.

Fig. 31 A schematic representation of the major plates involved in continental drift. The arrows show the direction of the present-day plate movements. The irregular offset pattern along some plate edges is due to transform faulting. In some areas the plate boundaries are not yet well defined (see also pp. 48–64).

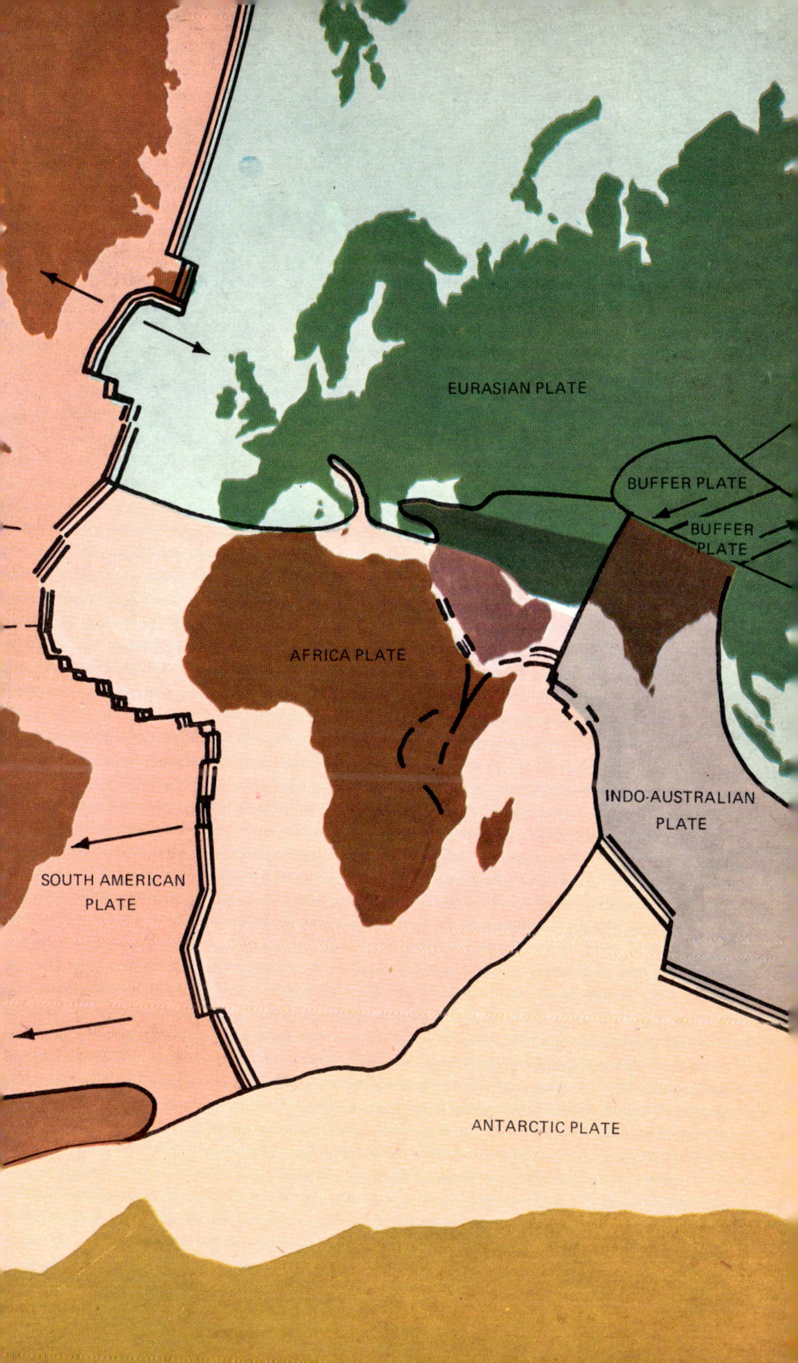

EURASIAN PLATE
BUFFER PLATE
BUFFER PLATE
AFRICA PLATE
INDO-AUSTRALIAN PLATE
SOUTH AMERICAN PLATE
ANTARCTIC PLATE

Fig. 32 H.M.S. *Beagle* (see p. 170).

Fig. 33 H.M.S. *Challenger* (see pp. 171–2).

Fig. 34 The *Glomar Challenger*, a research drill ship of the Deep Sea Drilling Project (DSDP), developed by the Scripps Institution of Oceanography (see also p. 163).

EUPHOTIC ZONE
MESOPELAGIC ZONE
BATHYPELAGIC ZONE
BENTHIC ZONE
1
2
3
4
7
8
9
10
11
12
13
28
29
30
31
33
34
35
36
38
39
40
43
44
45
47
49
50
51

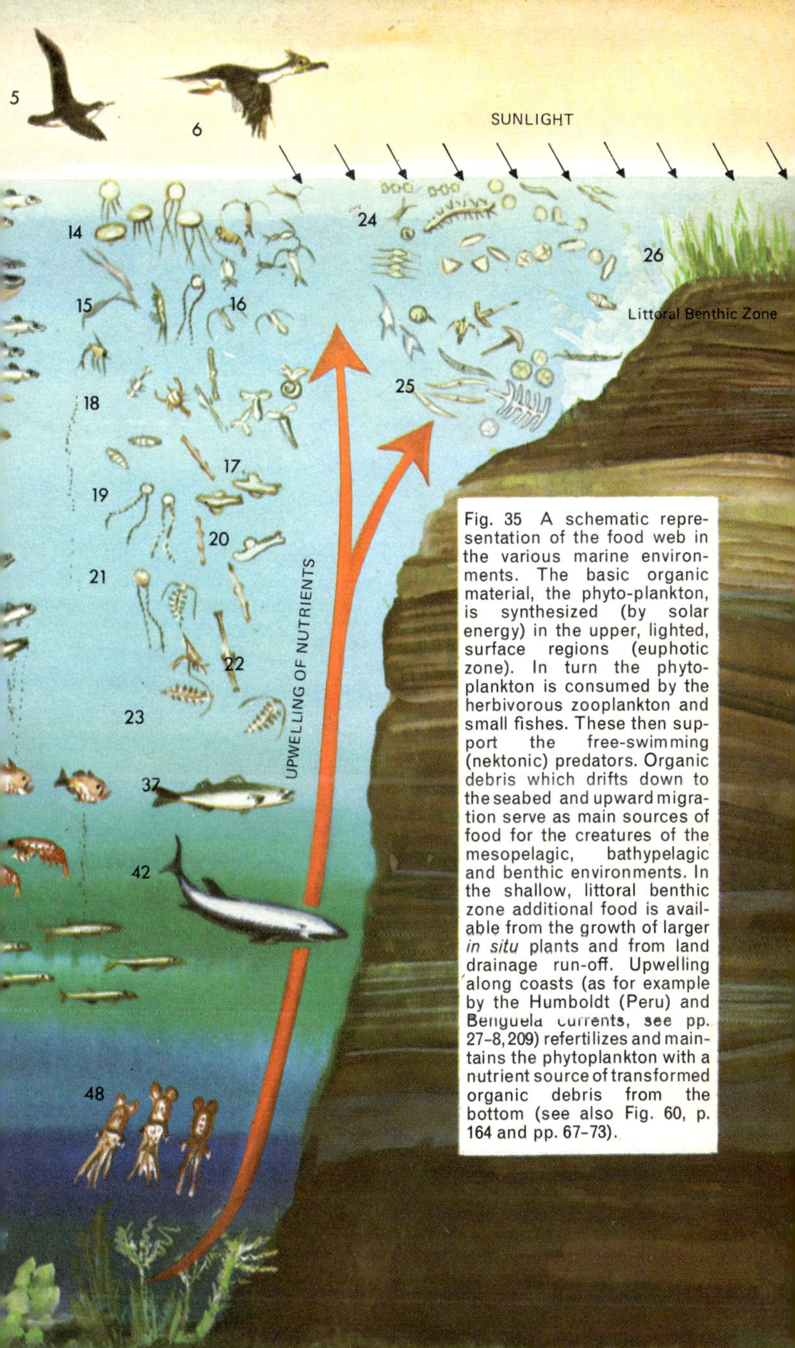

Fig. 35 A schematic representation of the food web in the various marine environments. The basic organic material, the phyto-plankton, is synthesized (by solar energy) in the upper, lighted, surface regions (euphotic zone). In turn the phytoplankton is consumed by the herbivorous zooplankton and small fishes. These then support the free-swimming (nektonic) predators. Organic debris which drifts down to the seabed and upward migration serve as main sources of food for the creatures of the mesopelagic, bathypelagic and benthic environments. In the shallow, littoral benthic zone additional food is available from the growth of larger *in situ* plants and from land drainage run-off. Upwelling along coasts (as for example by the Humboldt (Peru) and Benguela currents, see pp. 27–8, 209) refertilizes and maintains the phytoplankton with a nutrient source of transformed organic debris from the bottom (see also Fig. 60, p. 164 and pp. 67–73).

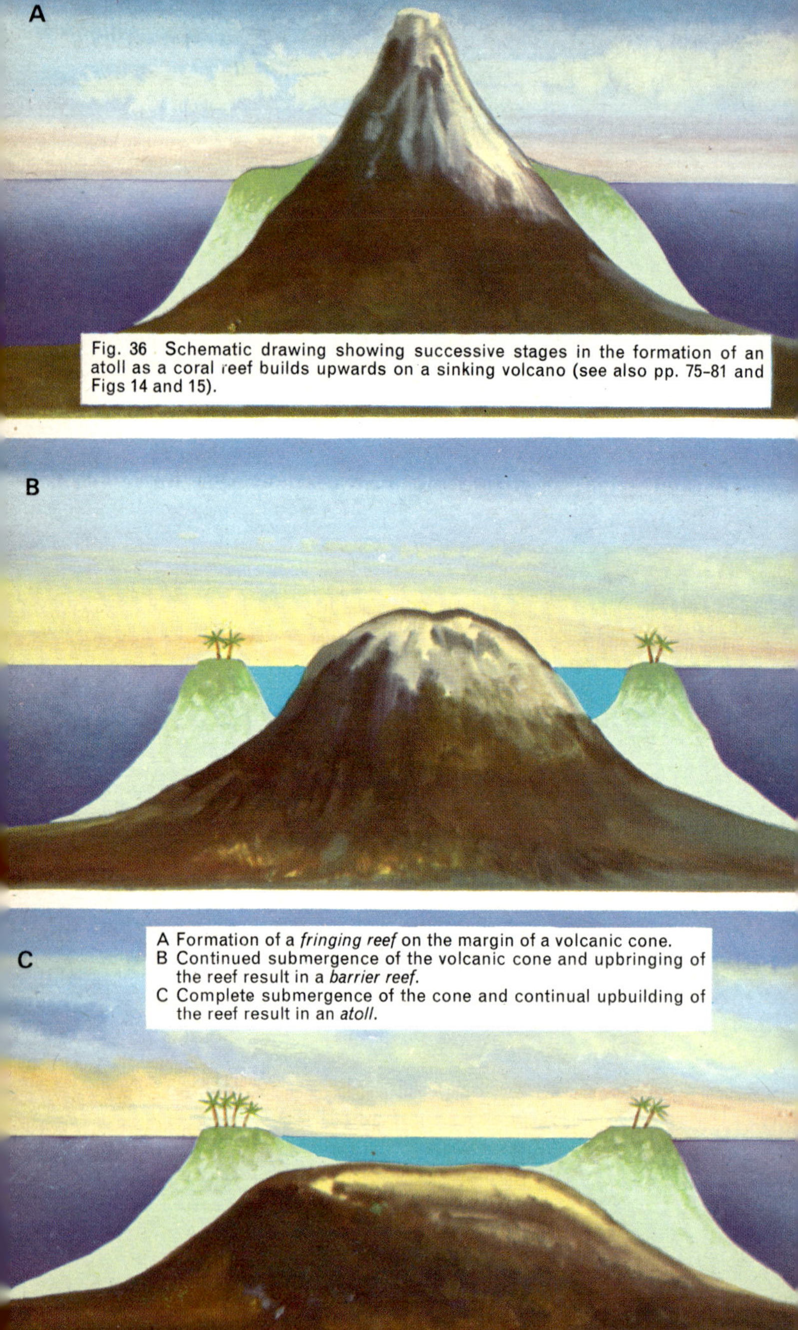

Fig. 36 Schematic drawing showing successive stages in the formation of an atoll as a coral reef builds upwards on a sinking volcano (see also pp. 75–81 and Figs 14 and 15).

A Formation of a *fringing reef* on the margin of a volcanic cone.
B Continued submergence of the volcanic cone and upbringing of the reef result in a *barrier reef*.
C Complete submergence of the cone and continual upbuilding of the reef result in an *atoll*.

ı. 37 The Crown of Thorns starfish (*Acanthaster planci*) which in more :ent years has caused concern owing to its ability to breed in large gregations and destroy substantial areas of coral reef in the Red Sea and Pacific and, in particular, on the Great Barrier Reef (see pp. 79–80).

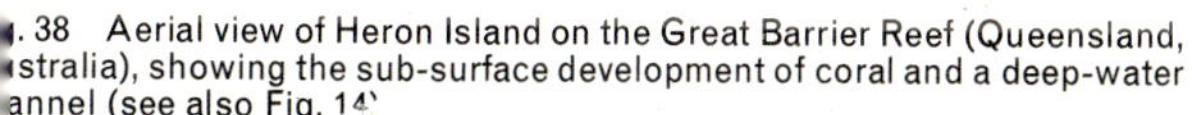

ı. 38 Aerial view of Heron Island on the Great Barrier Reef (Queensland, ıstralia), showing the sub-surface development of coral and a deep-water annel (see also Fig. 14)

Fig. 39 The bathyscape *Trieste* in which Jacques Piccard and Lt Walsh succeeded in descending to the bottom of the 10,800-metre (36,000-foot) Marianas Trench off Japan in January 1960 (see also p. 179). (*inset, top left*) The French bathyscape *Archimede* (appropriately named after the Greek who discovered the principle of buoyancy) also made a record dive, which was commemorated by the issue of a postage stamp in 1963.

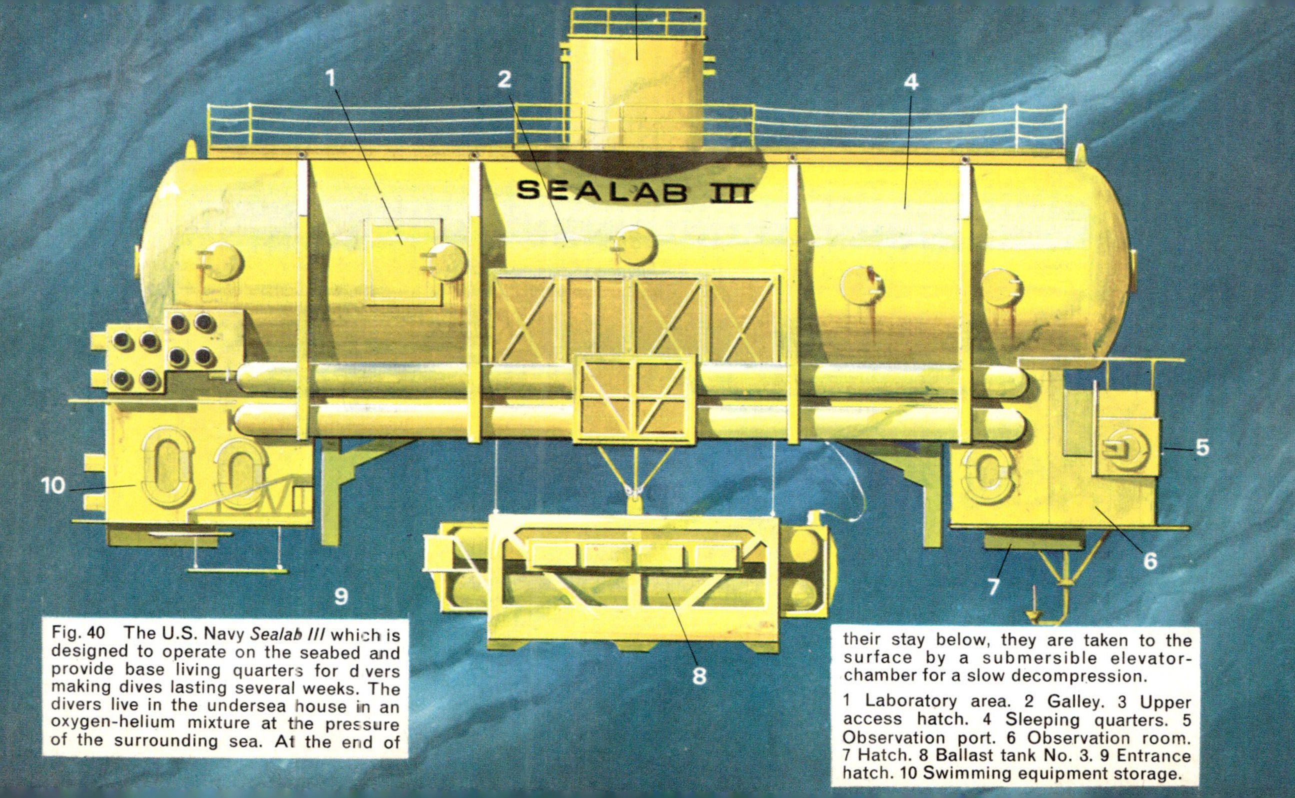

Fig. 40 The U.S. Navy *Sealab III* which is designed to operate on the seabed and provide base living quarters for divers making dives lasting several weeks. The divers live in the undersea house in an oxygen-helium mixture at the pressure of the surrounding sea. At the end of their stay below, they are taken to the surface by a submersible elevator-chamber for a slow decompression.

1 Laboratory area. 2 Galley. 3 Upper access hatch. 4 Sleeping quarters. 5 Observation port. 6 Observation room. 7 Hatch. 8 Ballast tank No. 3. 9 Entrance hatch. 10 Swimming equipment storage.

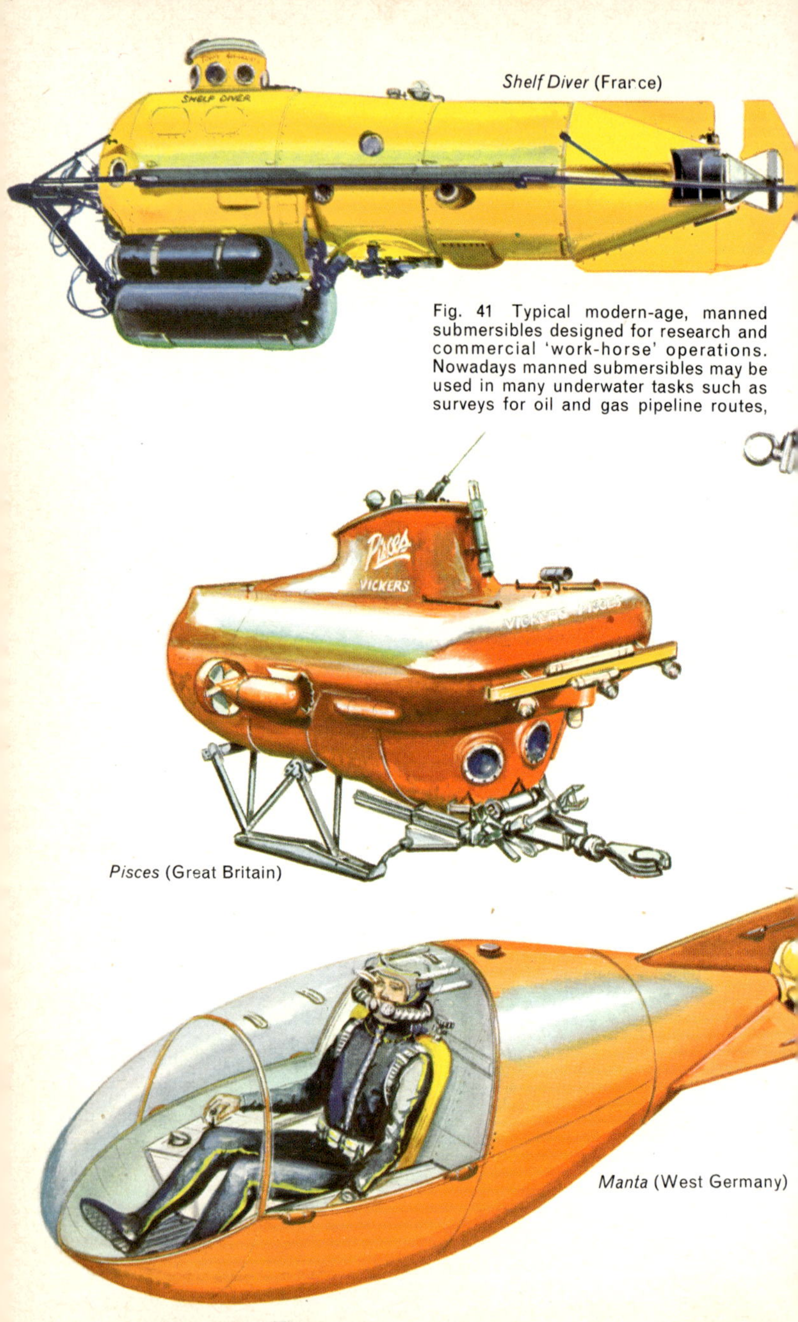

Fig. 41 Typical modern-age, manned submersibles designed for research and commercial 'work-horse' operations. Nowadays manned submersibles may be used in many underwater tasks such as surveys for oil and gas pipeline routes,

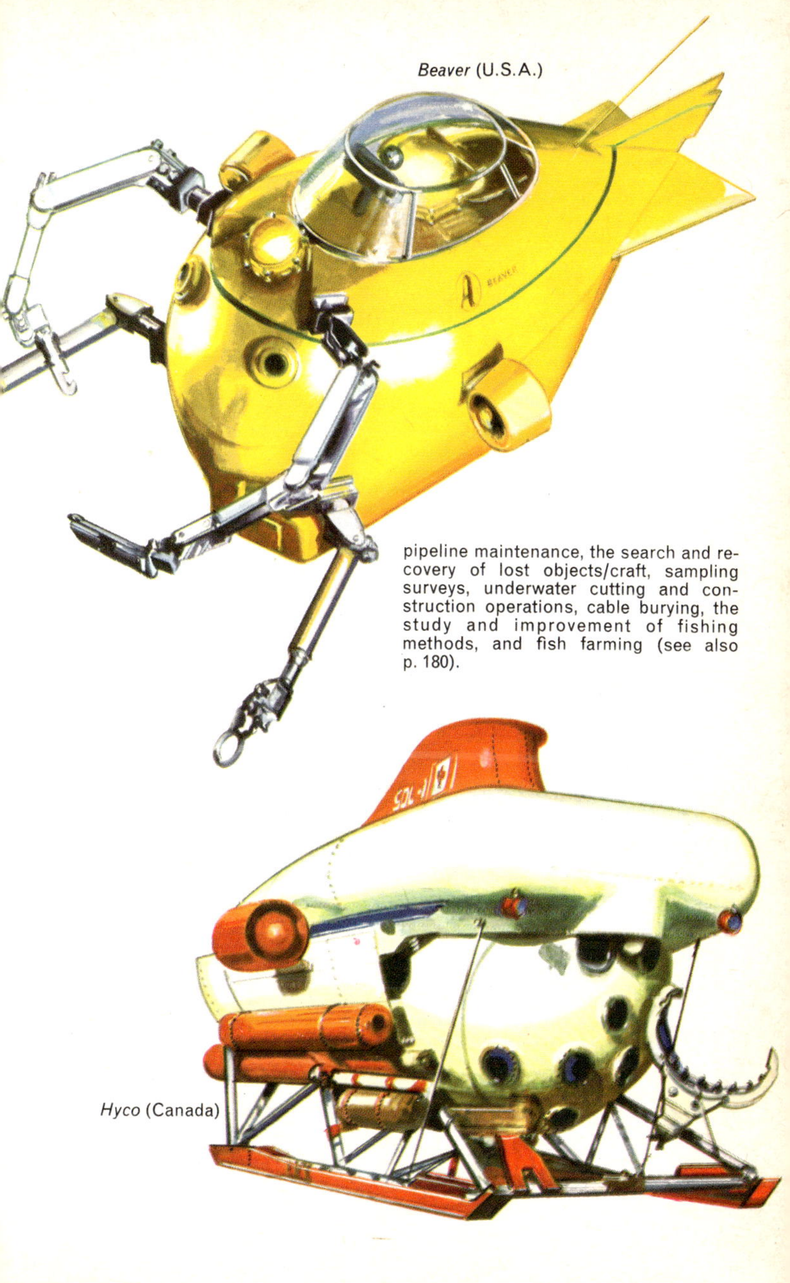

Beaver (U.S.A.)

pipeline maintenance, the search and recovery of lost objects/craft, sampling surveys, underwater cutting and construction operations, cable burying, the study and improvement of fishing methods, and fish farming (see also p. 180).

Hyco (Canada)

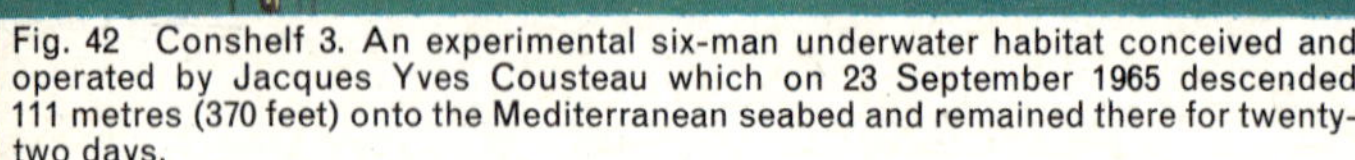

Fig. 42 Conshelf 3. An experimental six-man underwater habitat conceived and operated by Jacques Yves Cousteau which on 23 September 1965 descended 111 metres (370 feet) onto the Mediterranean seabed and remained there for twenty-two days.

1 Neoprene net covered bag containing 9 tons of wash water. 2 Mesh pantry box (outside 'refrigerator'). 3 Sleeping quarters. 4 IBM teleprinter. 5 Telephone and radio equipment. 6 Mess. 7 Infra-red oven/cooker. 8 No. 1 lifeboat and decompression chamber. 9 Galley. 10 Sink. 11 Entrance hatch. 12 5·4 metre (18 foot) diameter sphere. 13 No. 2 lifeboat and decompression chamber. 14 Helium and oxygen circulating cylinder (which also serves as a deep freeze). 15 Companion-way to lower deck. 16 Deck. 17 Ballast tank (filled with 77 tons of ballast). 18 Diving hatch. 19 Navigation/identification lights. 20 W.C. 21 Foot. 22 Chassis 14·4 × 8·4 metres (48 × 28 feet). 23 Oil well-head. 24 Diving saucer. 25 Aqualung pack plus one reserve pack. 26 Oceanaut in foam rubber wet suit.

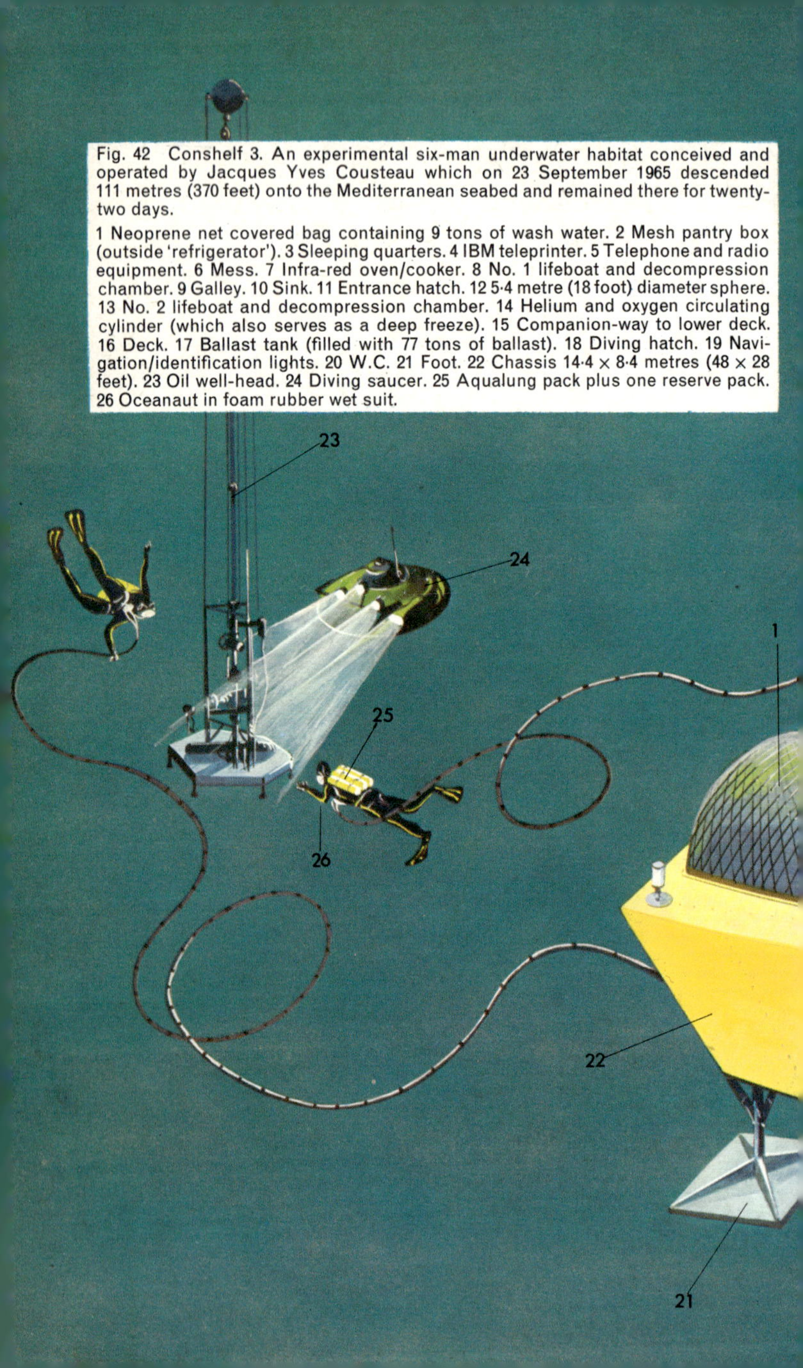

8
9
11
10
12
13
7
6
14
15
16
17
18
19
20

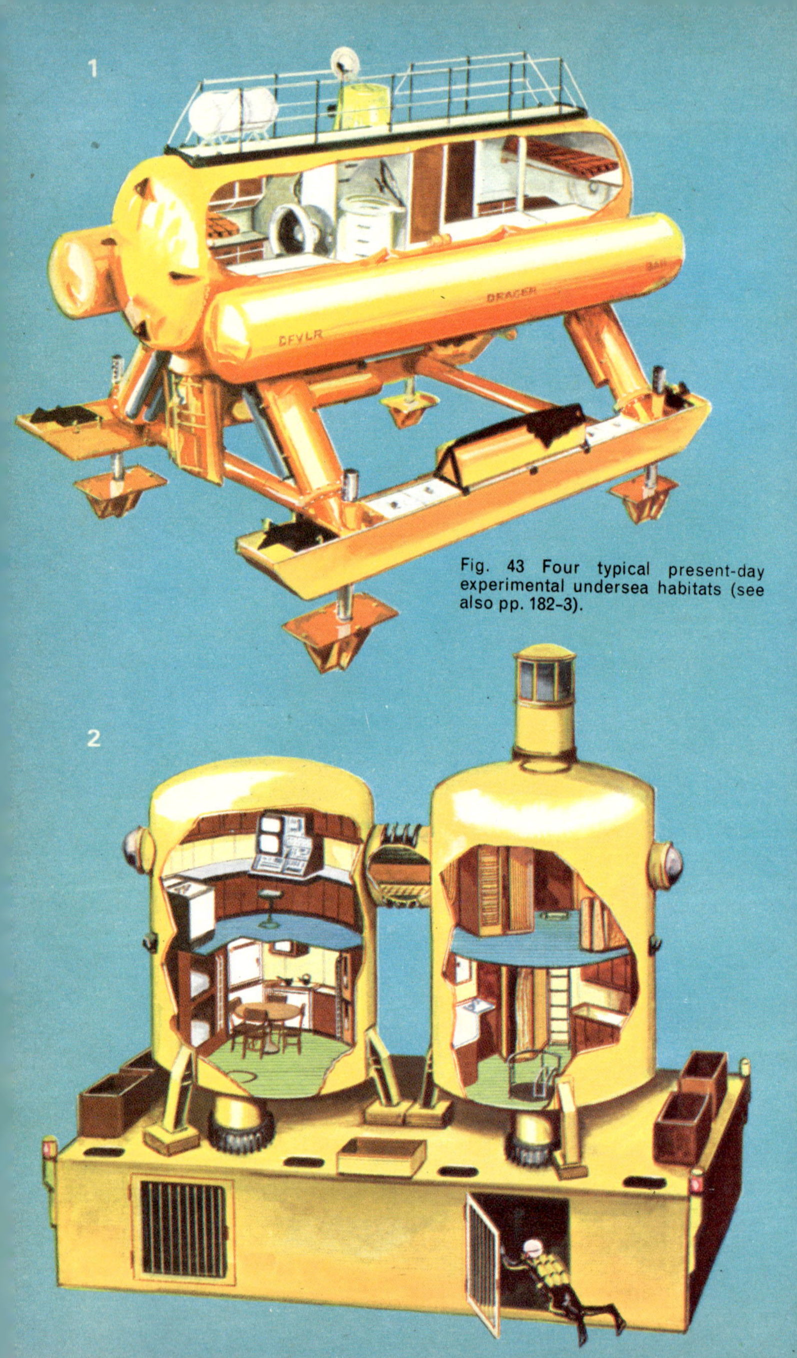

Fig. 43 Four typical present-day experimental undersea habitats (see also pp. 182–3).

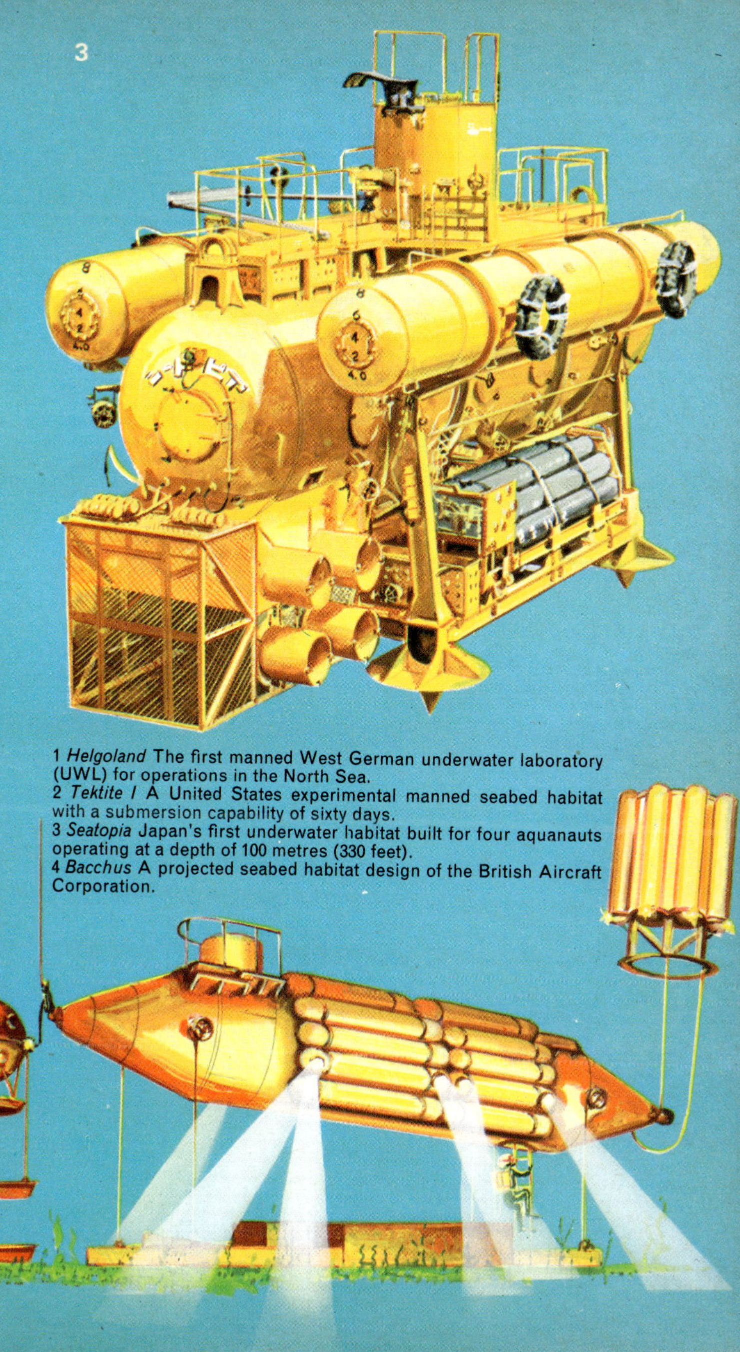

1 *Helgoland* The first manned West German underwater laboratory (UWL) for operations in the North Sea.
2 *Tektite I* A United States experimental manned seabed habitat with a submersion capability of sixty days.
3 *Seatopia* Japan's first underwater habitat built for four aquanauts operating at a depth of 100 metres (330 feet).
4 *Bacchus* A projected seabed habitat design of the British Aircraft Corporation.

Fig. 44 Floating Instrument Platform (FLIP) developed by the Scripps Institution of Oceanography. The vessel is 107 metres (355 feet) long and operates with a crew of three. It is first towed to the operations site, and once in position, it is flooded so that it upends and sits vertically in the water and becomes a stable platform resting firmly on the bottom. One of its functions has been to measure open-sea wave heights in the Pacific Ocean. A wave-measuring instrument is located on the sea bottom and operates by recording fluctuations of water pressure above it. These signals are recorded on magnetic tape, then onto a punched paper tape for direct input into the computer at the Scripps Institution in California.

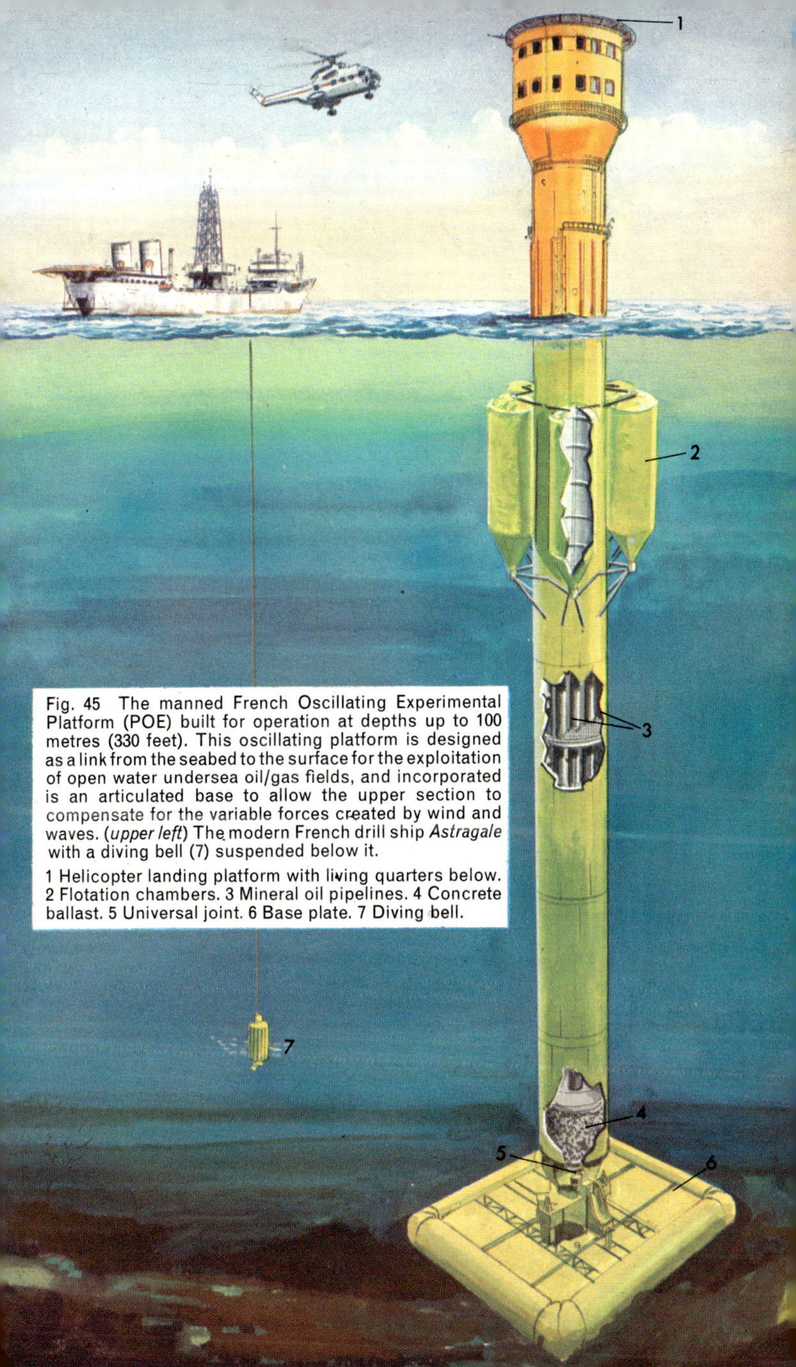

Fig. 45 The manned French Oscillating Experimental Platform (POE) built for operation at depths up to 100 metres (330 feet). This oscillating platform is designed as a link from the seabed to the surface for the exploitation of open water undersea oil/gas fields, and incorporated is an articulated base to allow the upper section to compensate for the variable forces created by wind and waves. (*upper left*) The modern French drill ship *Astragale* with a diving bell (7) suspended below it.

1 Helicopter landing platform with living quarters below. 2 Flotation chambers. 3 Mineral oil pipelines. 4 Concrete ballast. 5 Universal joint. 6 Base plate. 7 Diving bell.

Fig. 46 *Ocean Prospector*, one of the world's largest semi-submersible, self-propelled drilling platforms; length 104·3 metres (342·3 feet); breadth 80·3 metres (263·5 feet); height 38·4 metres (126·0 feet). This vessel can operate in depths up to 200 metres (660 feet) and has a drilling capability of 8,000 metres (26,400 feet). (*inset* A) A view of the vessel showing the two propulsion units (2×2,700 h.p.). (*inset* B) A schematic interpretation of how a drilling platform (or rig) may have a capability of drilling multiple holes (up to *c.* 20) in the seabed in the arrangement of a fan from a single station.

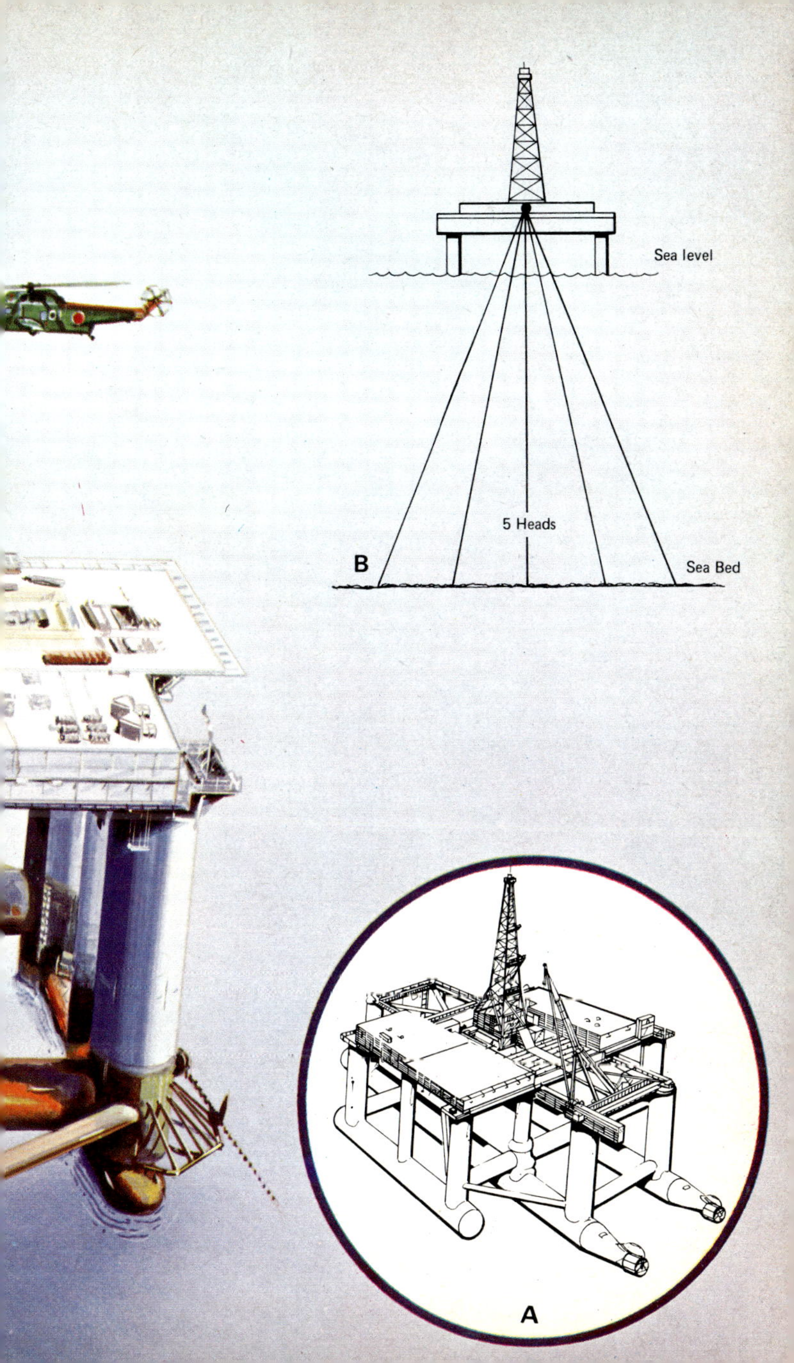
Sea level
5 Heads
B
Sea Bed
A

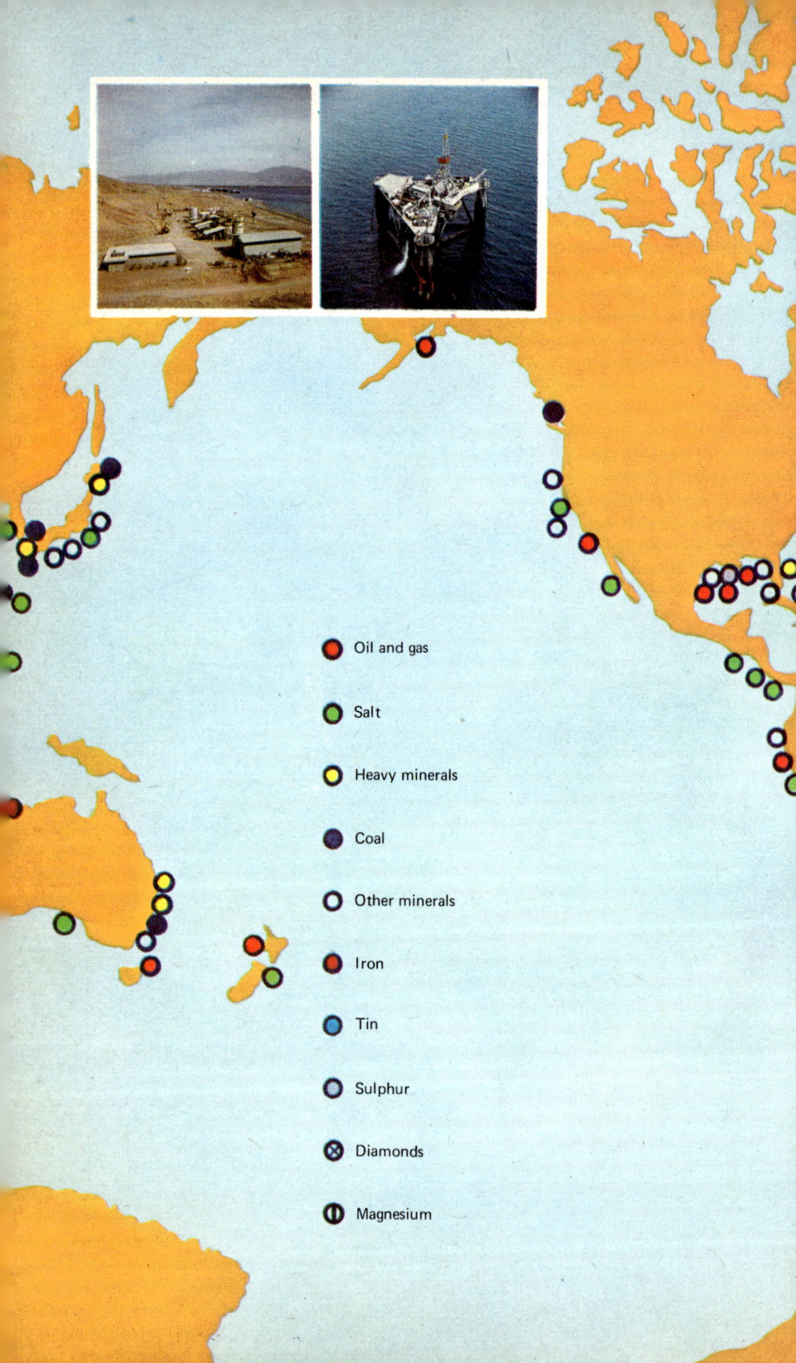
Oil and gas
Salt
Heavy minerals
Coal
Other minerals
Iron
Tin
Sulphur
Diamonds
Magnesium

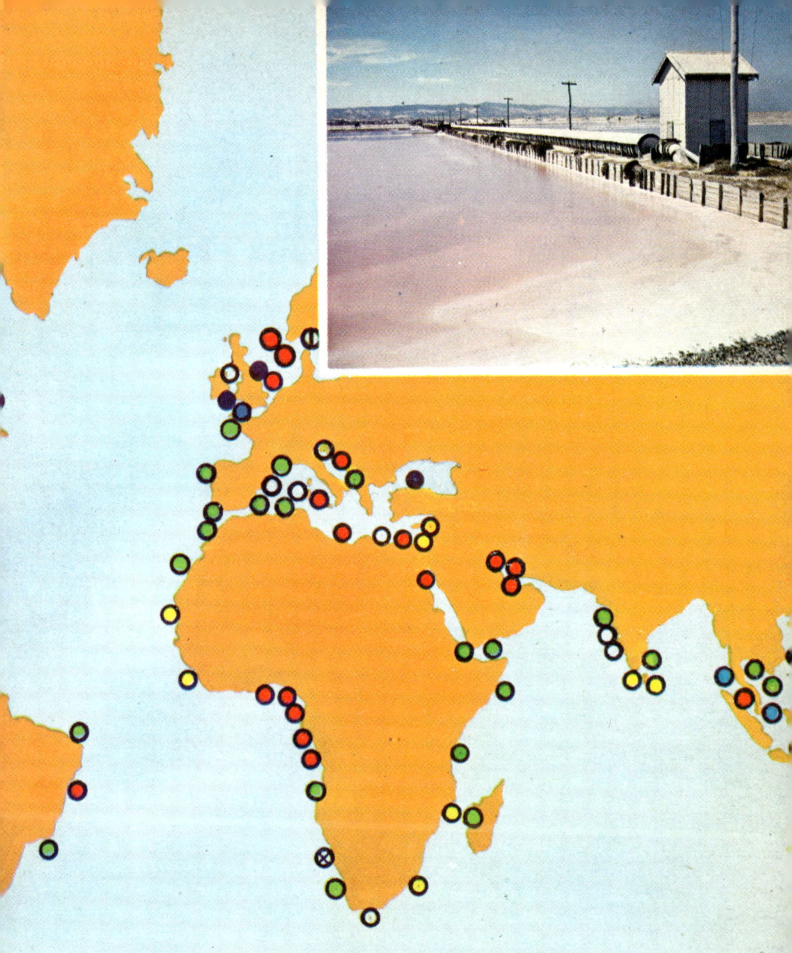

Fig. 47 Locations for major coastal and offshore mineral resources. At the present time resources being utilized range from shallow beach deposits to deposits on or under the deeper parts of the continental shelves.
(*inset, right*) In areas of high evaporation, solar salt farming is now a major coastal industry such as in Evaporation Bay, South Australia.
(*inset, far left*) Inland seas are also areas containing valuable resources such as potash currently in production on the shores of the Dead Sea.
(*inset, left*) Two of the most valuable strategic resources of the continental shelf are oil and gas, and throughout the world drilling platforms are tapping vast quantities of these fossil fuels.

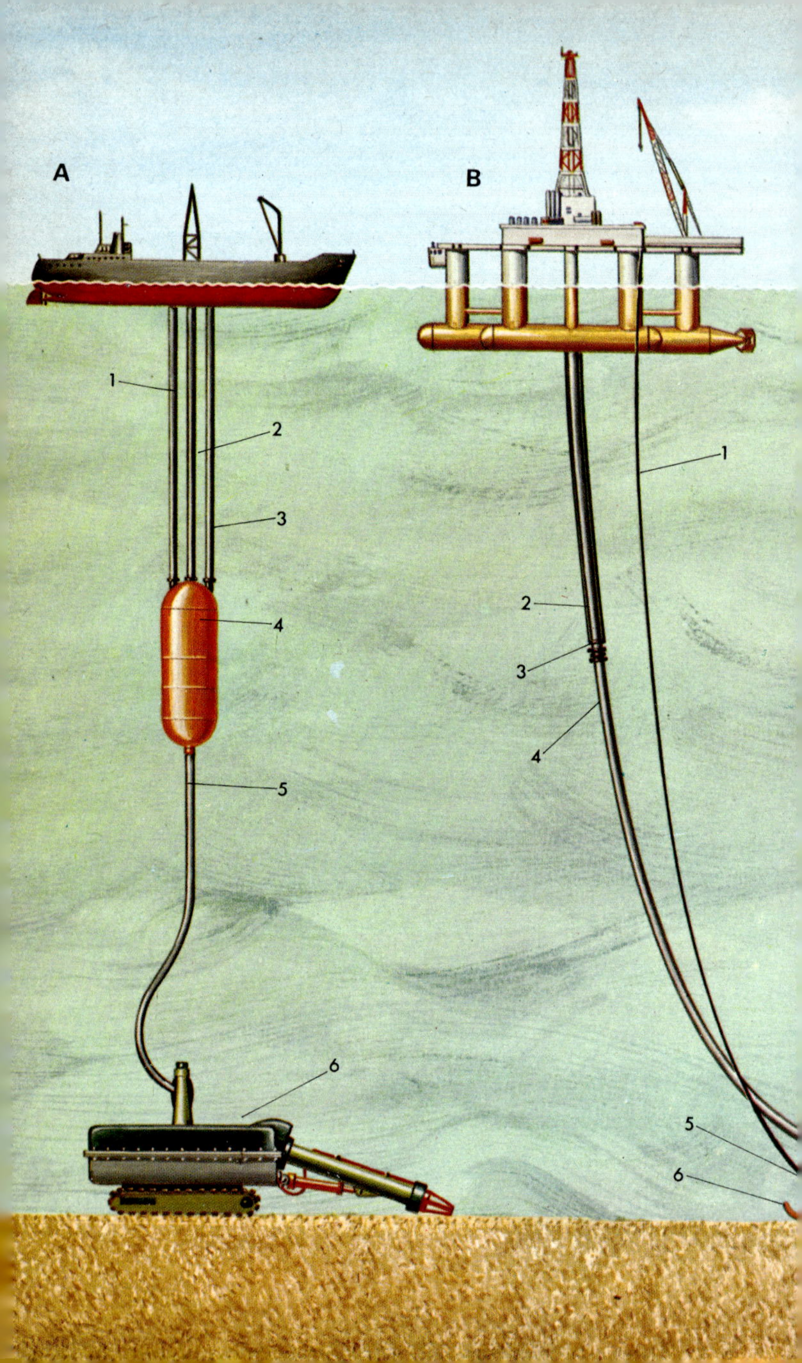
A
B
1
2
3
4
5
6
1
2
3
4
5
6

Fig. 48 Various ideas and schemes have been developed for mining the valuable minerals of the seabed (see also p. 164 and pp. 185–6).

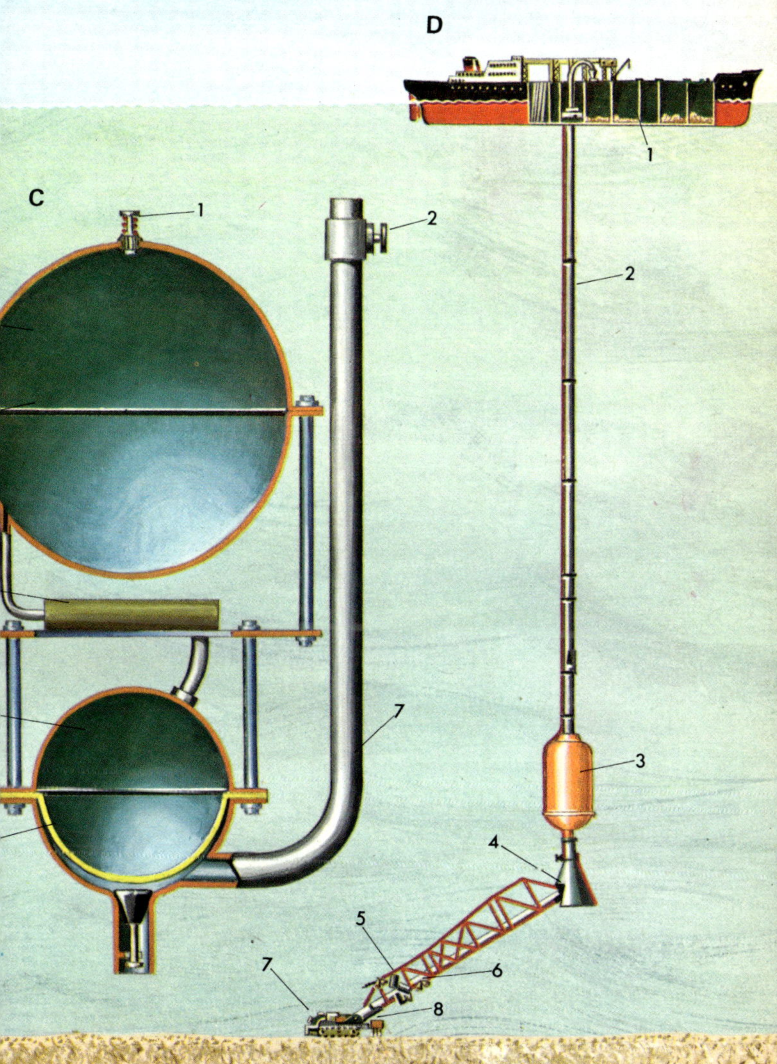

Figs 49 and 50 Examples of modern seabed technological aids. Fig. 49 A French-designed, ballasted submarine drilling rig powered by a surface diesel generator and operated by divers. In water depths to 50 metres (165 feet) it can drill holes 1·2 metres (4 feet) in diameter and *c*. 20 metres (65 feet) deep. Fig. 50 Underwater pipeline welding shop designed to provide full seabed protection for oil/gas line welding crews.

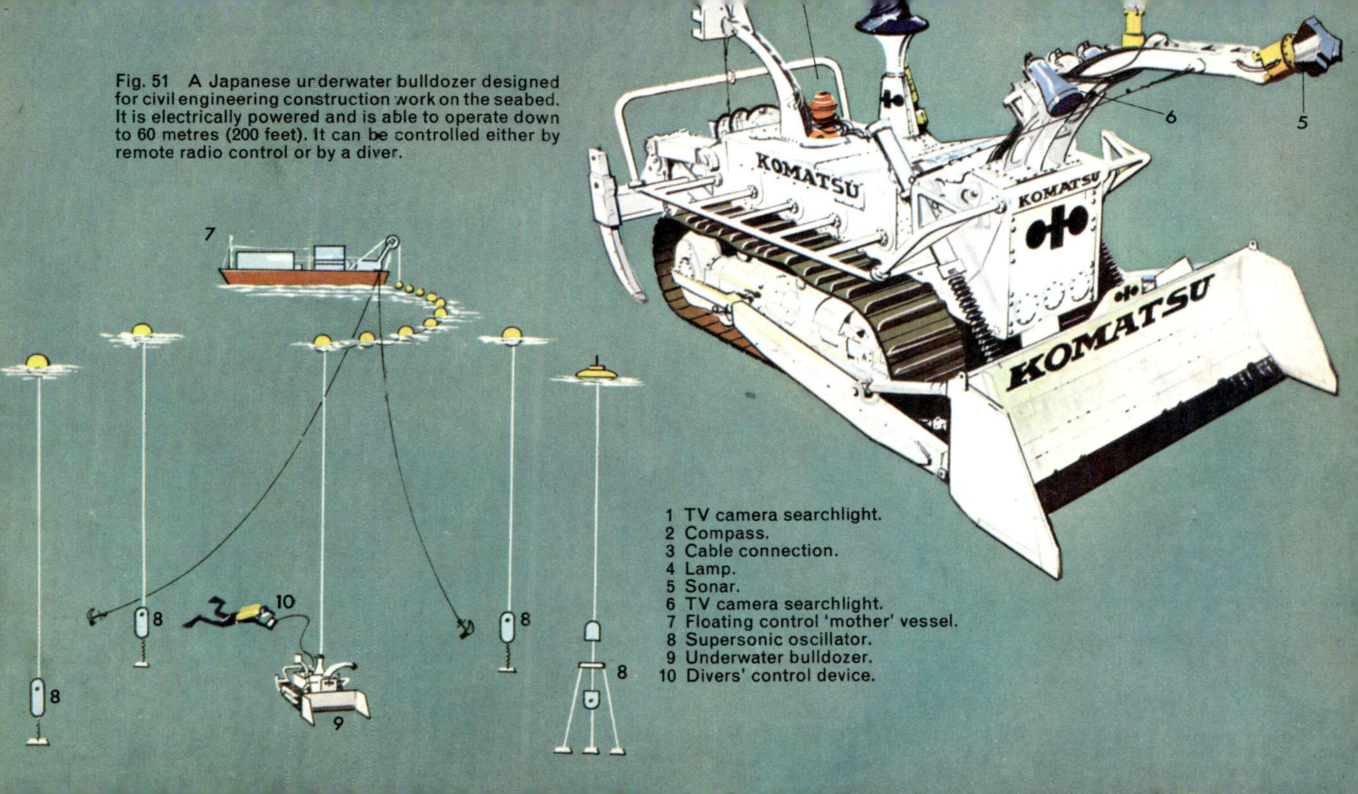

Fig. 51 A Japanese underwater bulldozer designed for civil engineering construction work on the seabed. It is electrically powered and is able to operate down to 60 metres (200 feet). It can be controlled either by remote radio control or by a diver.

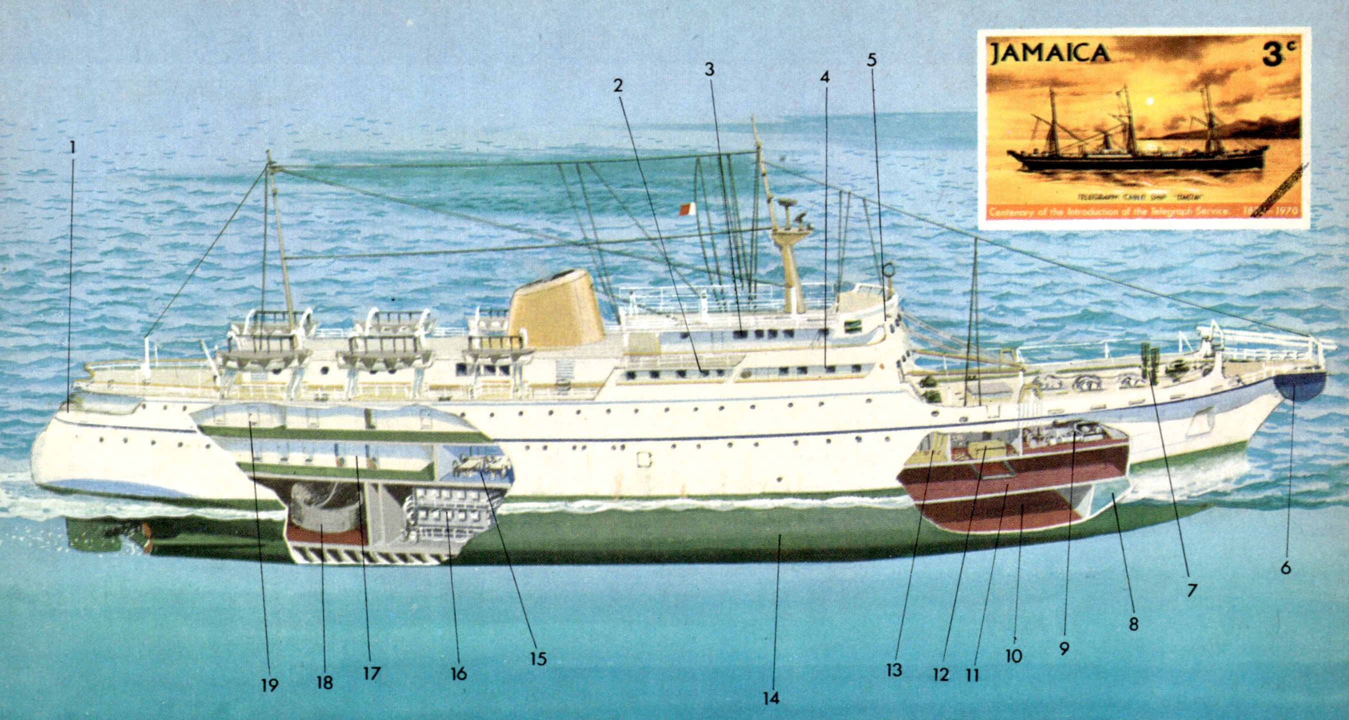

Fig. 52 Cable ships are continually in operation maintaining or laying new cables on the seabed. The 4,200-ton *Retriever* is a typical modern cable ship (see also p. 165). (*inset right*) The nineteenth-century cable ship *Dacia*

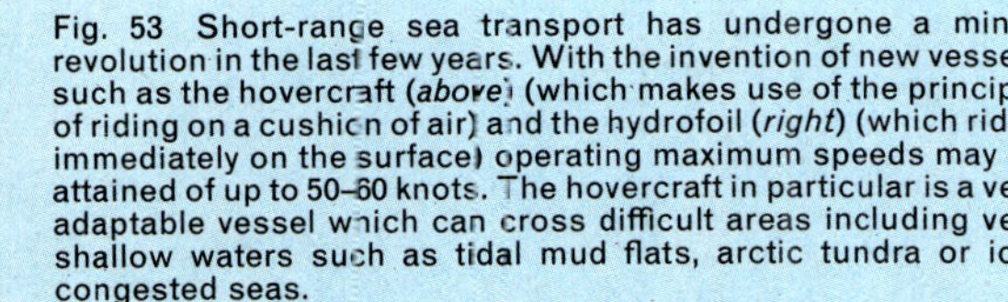

Fig. 53 Short-range sea transport has undergone a minor revolution in the last few years. With the invention of new vessels such as the hovercraft (*above*) (which makes use of the principle of riding on a cushion of air) and the hydrofoil (*right*) (which rides immediately on the surface) operating maximum speeds may be attained of up to 50–60 knots. The hovercraft in particular is a very adaptable vessel which can cross difficult areas including very shallow waters such as tidal mud flats, arctic tundra or ice-congested seas.

Fig. 54 Stratification of ocean waters. A thin warm layer overlies progressively colder water, with a sharp break at about 150 metres (500 feet). The surface water is denuded of nutrients but these are increasingly plentiful in deeper water (see pp. 207–8).

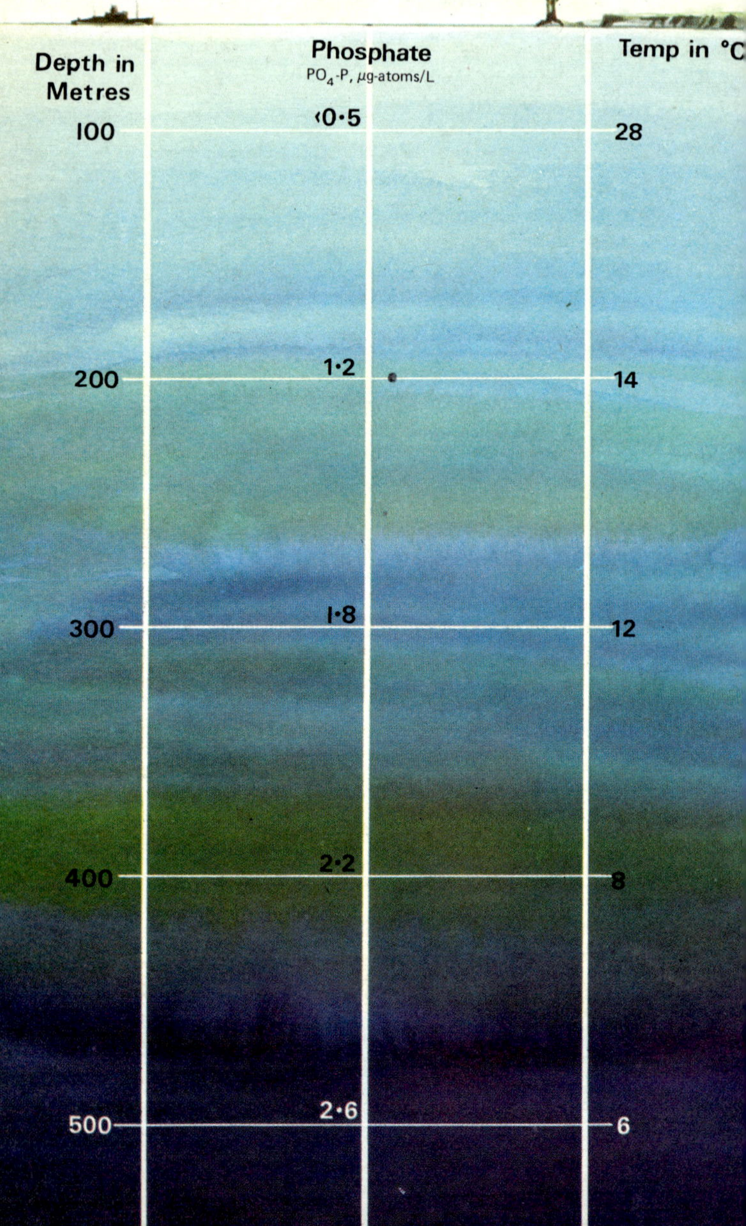

Fig. 55 Means by which nutrient-rich deeper water is lifted to the surface: (*top*) surface winds blow surface water offshore, deep water upwells to replace it; (*bottom*) where currents diverge, deeper water is dragged up (see pp. 209–10).

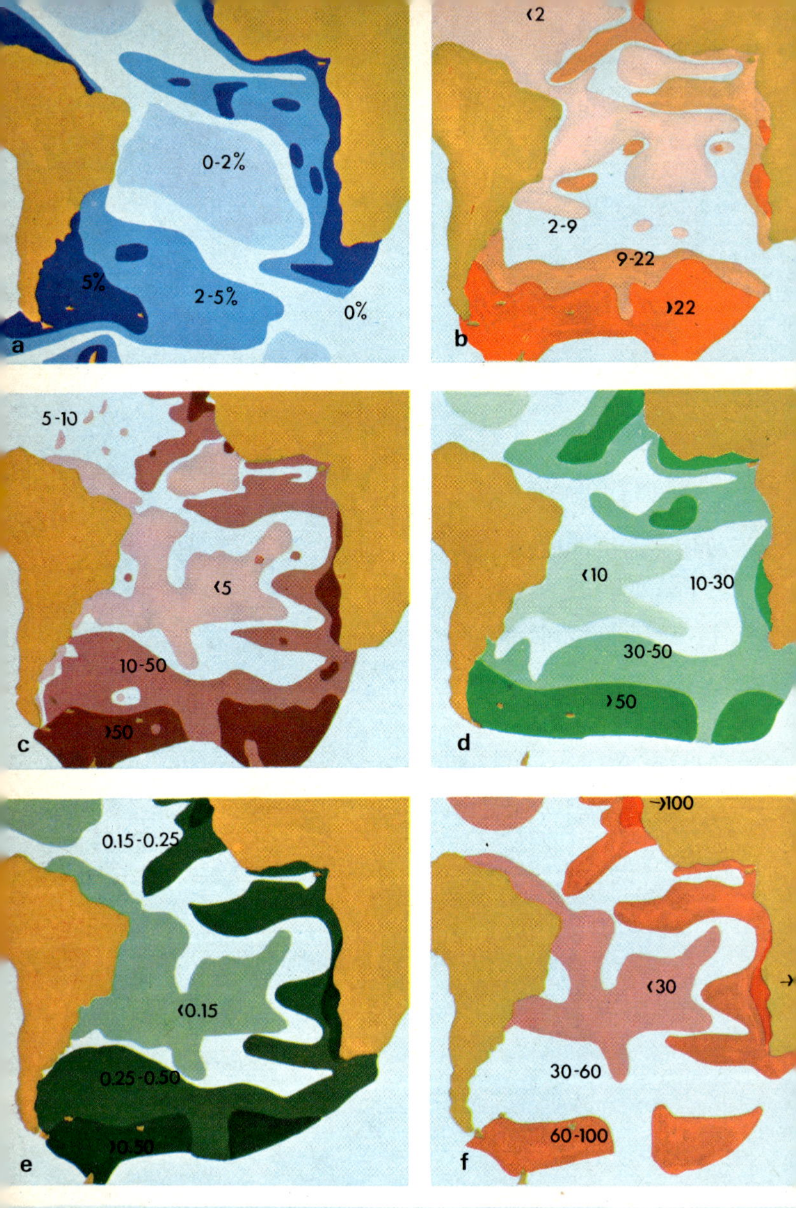

Fig. 56 Upwelling of nutrient-rich deeper water on the west coast of Africa (see p. 209).
a Colour of the ocean: 0–2=deep blue, 2–5=blue-green, 5=green. b Distribution of phosphate, milligrams per cubic metre. c Distribution of plankton organisms, thousands/litre. d Distribution of zooplankton, numbers per 4 litres. e Distribution of organic gross production in summer, gC/m²/day (see p. 209). f Distribution of annual net production, gC/m²/year.

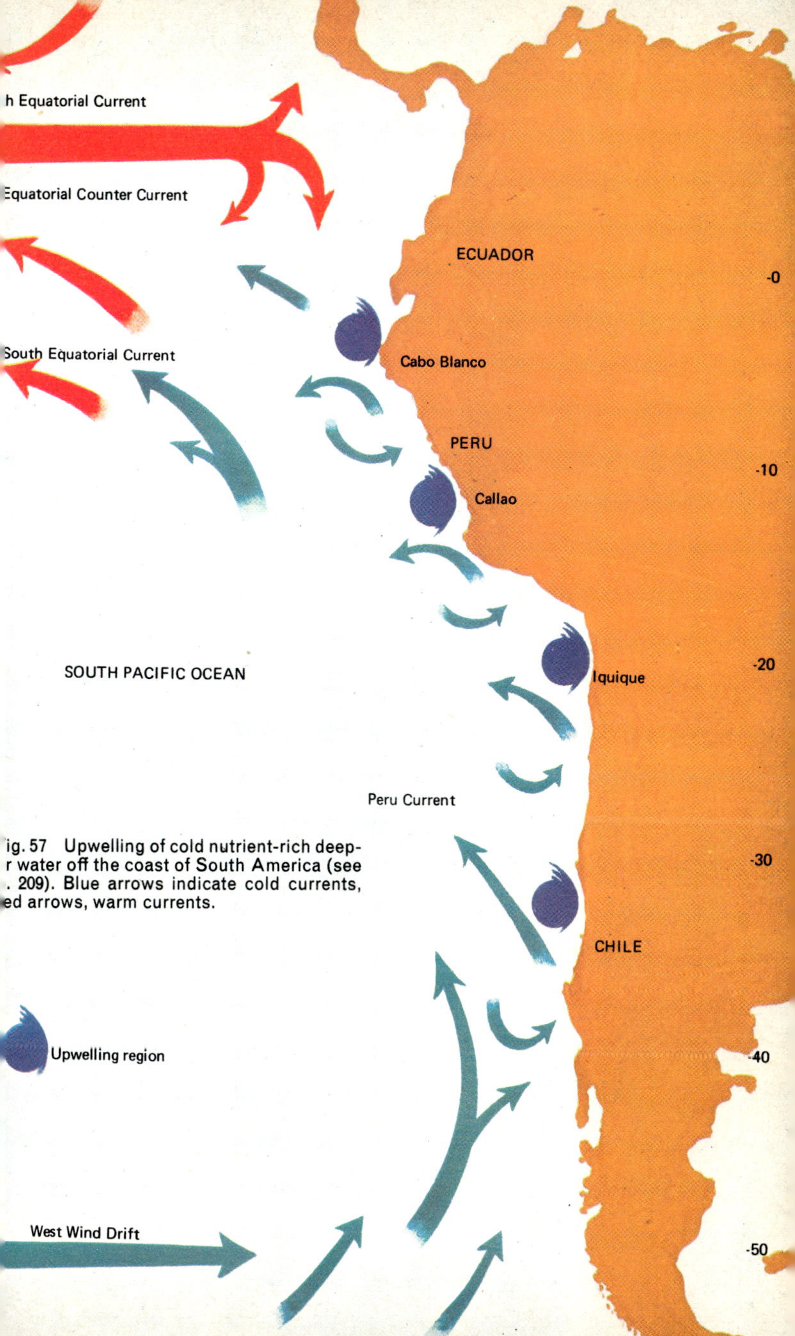

ig. 57 Upwelling of cold nutrient-rich deep-r water off the coast of South America (see . 209). Blue arrows indicate cold currents, ed arrows, warm currents.

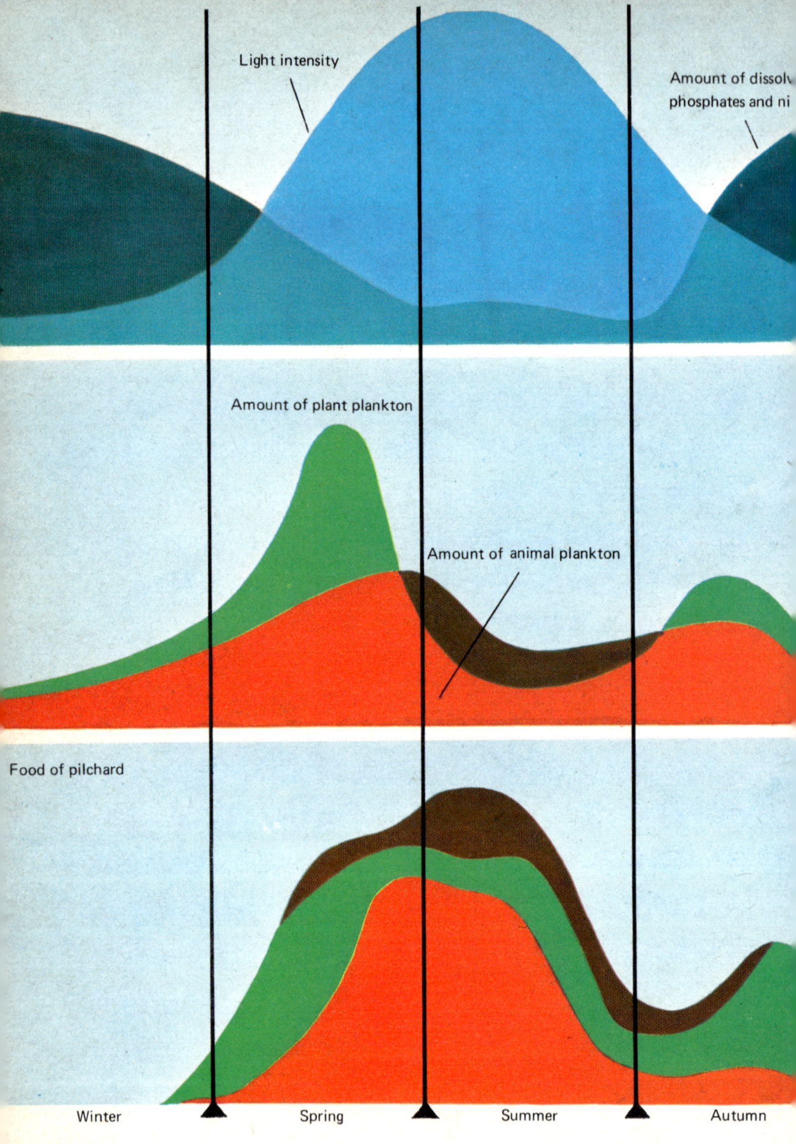

Fig. 58 Seasonal variation in plankton production in the English Channel. This is reflected in the food and feeding of the Cornish pilchard. (*top curves*) Light intensity and amount of dissolved nutrient salts. (*middle curves*) Production of phytoplankton, in green, and zooplankton, in red. (*bottom curves*) Stomach contents of pilchard, red for copepods, green for phytoplankton, buff for other zooplankton (see pp. 207–9).

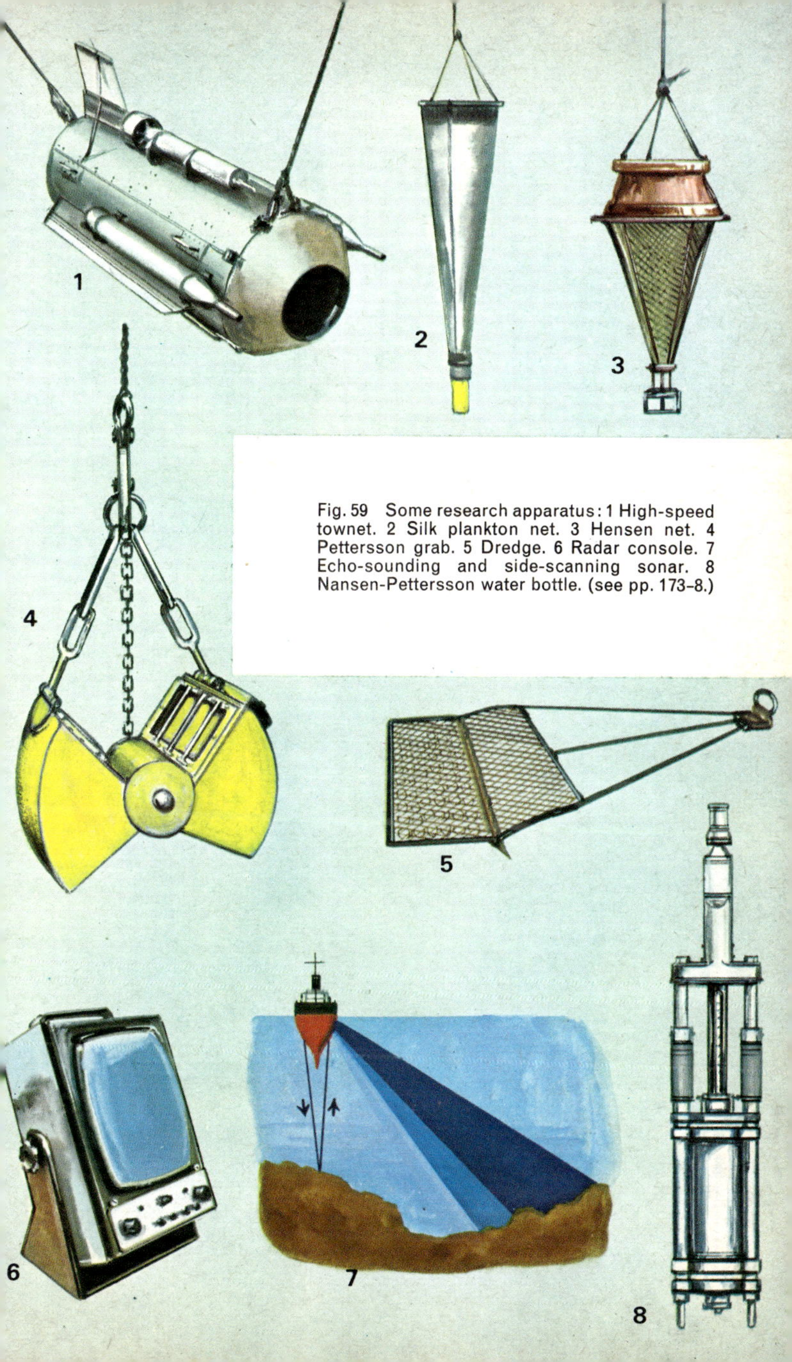

Fig. 59 Some research apparatus: 1 High-speed townet. 2 Silk plankton net. 3 Hensen net. 4 Pettersson grab. 5 Dredge. 6 Radar console. 7 Echo-sounding and side-scanning sonar. 8 Nansen-Pettersson water bottle. (see pp. 173–8.)

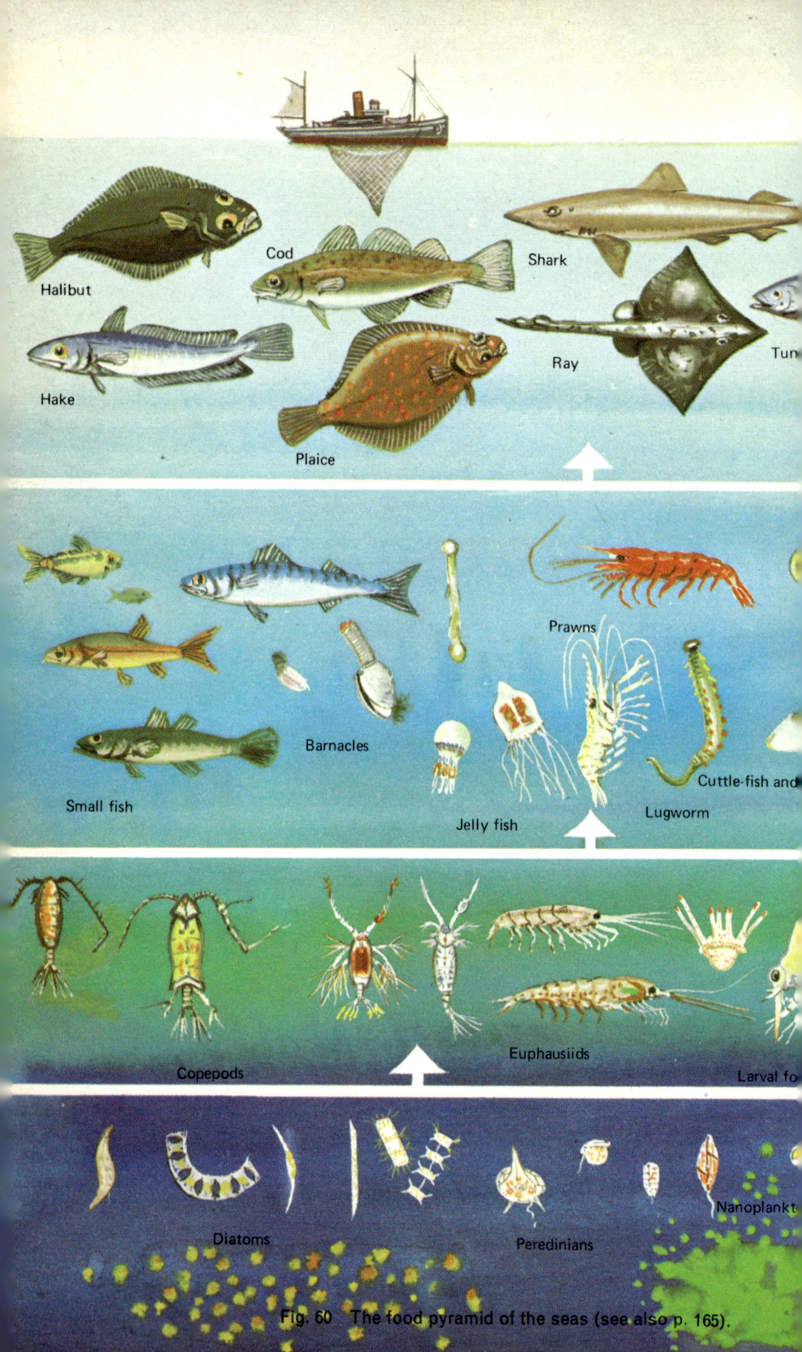

Fig. 60 The food pyramid of the seas (see also p. 165).

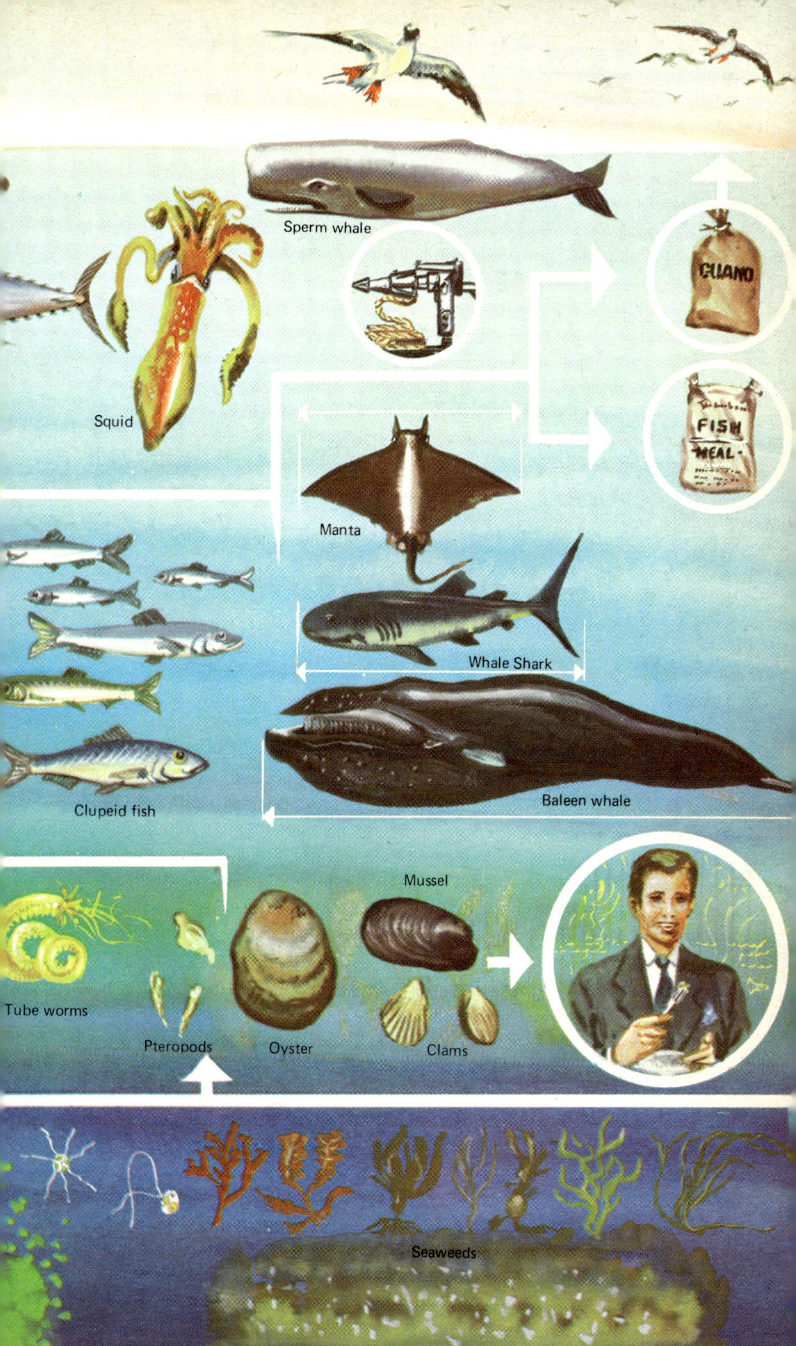
Sperm whale
GUANO
FISH
MEAL
Squid
Manta
Whale Shark
Clupeid fish
Baleen whale
Mussel
Tube worms
Pteropods
Oyster
Clams
Seaweeds

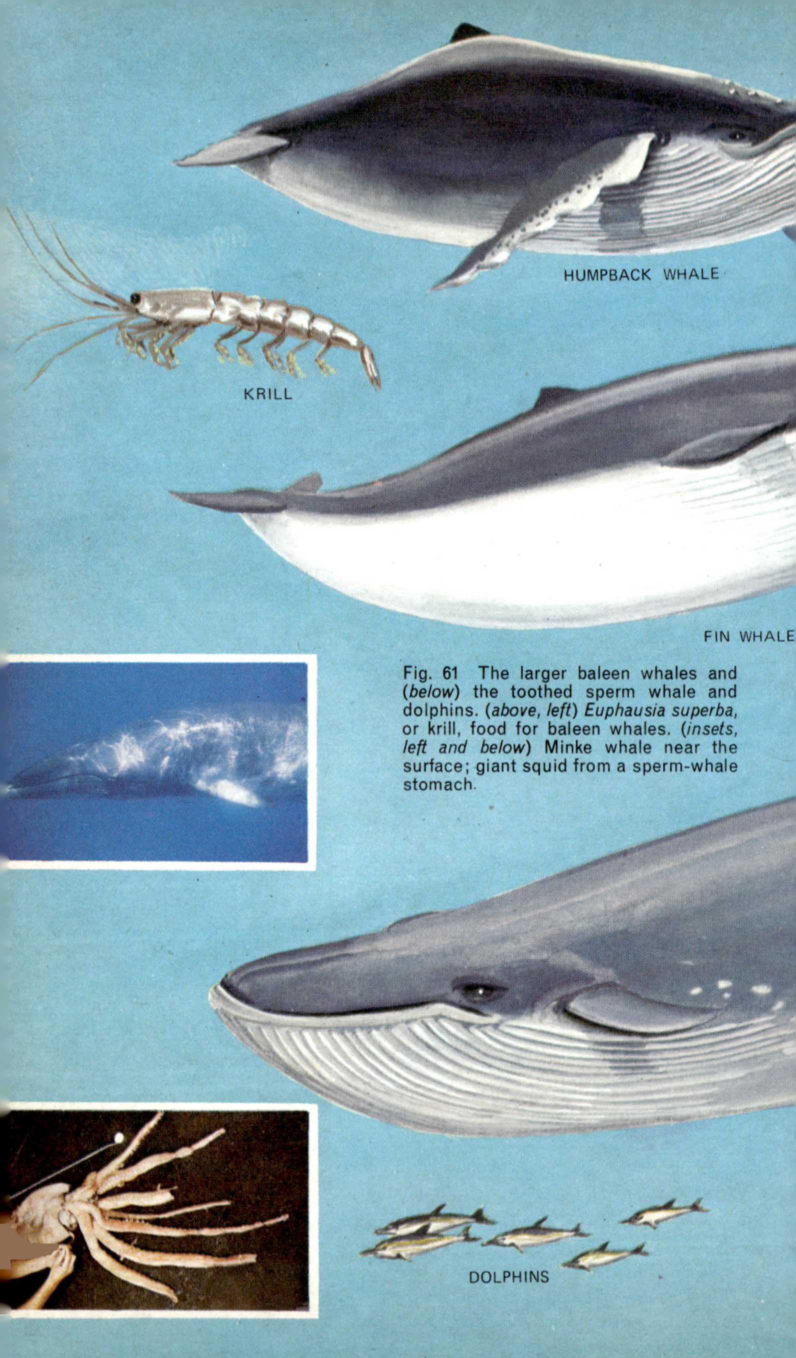

Fig. 61 The larger baleen whales and (*below*) the toothed sperm whale and dolphins. (*above, left*) *Euphausia superba*, or krill, food for baleen whales. (*insets, left and below*) Minke whale near the surface; giant squid from a sperm-whale stomach.

RIGHT WHALE
SEI WHALE
WHALE
SPERM WHALE

Fig. 62 Animal life on the abyssal plains. 1–3 Bristle-worms. 4 Sea cucumber, *Scotoplanes*. 5 *Limopsis*. 6 *Eremicaster*, a starfish. 7 *Xylophaga*. 8 *Neopilina*. 9 Giant hydroid. 10 Hermit crab. 11 Amphipod. 12 Rat-tail, *Macrouroides*. 13 *Bassogigas*, one of the deepest-dwelling fish. 14 Tripod fish, *Benthosaurus*. 15 The unique *Galathaeathauma*, with a luminescent lure inside the mouth.

Fig. 63 Deep-sea pelagic life (see p. 212): 1 Mysid. 2 Arrow-worm. 3 and 4 Copepods. 5 *Vampyroteuthis*, a deep-sea octopus. 6 Deep-sea squid. 7 Lantern-fish. 8 Bristle-mouth. 9 Deep-sea prawn. 10 Moleskin shark. 11 Deep-sea shark. 12 Bob-tailed snipe-eel. 13 Deep-sea angler-fish. 14 Whale-fish.

Fig. 64 Species of fish which are probably under-exploited (see pp. 229–30): 1 Silver hake. 2 Blue whiting. 3 Anchovies. 4 Sardinellas. 5 Bristlemouths. 6 Grenadier fish.

Fig. 65 Soviet research ship *Akaoemik Knipovitch* (see p. 228). (*below*) An experim ental Sov et plankton harvester.

1 Net.
2 Net frame with sinker.
3 Suction.
4 Bridles.
5 Swing boom.
6 Guy wire.
7 Water separator.
8 Pump.

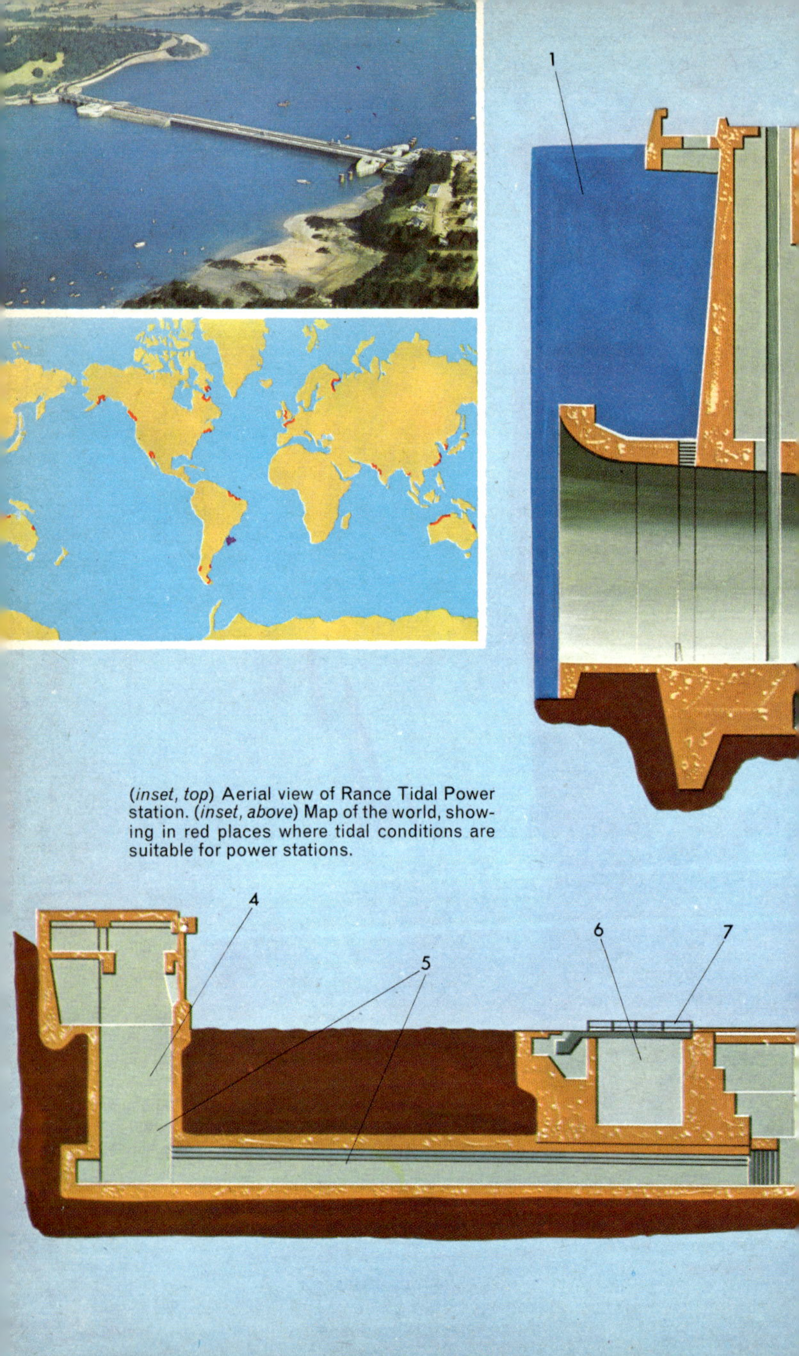

(*inset, top*) Aerial view of Rance Tidal Power station. (*inset, above*) Map of the world, showing in red places where tidal conditions are suitable for power stations.

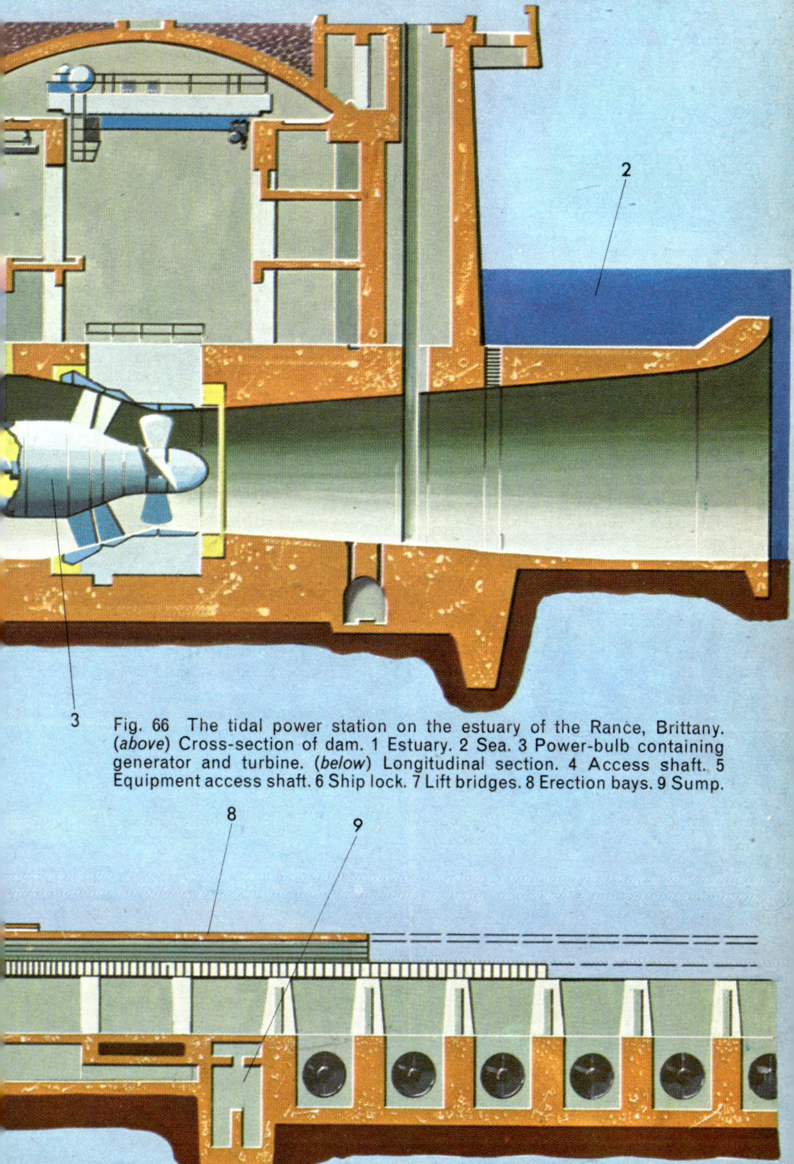

2

3

Fig. 66 The tidal power station on the estuary of the Rance, Brittany. (*above*) Cross-section of dam. 1 Estuary. 2 Sea. 3 Power-bulb containing generator and turbine. (*below*) Longitudinal section. 4 Access shaft. 5 Equipment access shaft. 6 Ship lock. 7 Lift bridges. 8 Erection bays. 9 Sump.

Fig. 67 A modern design for a tidal turbine (see p. 198). 1 Sluice gate. 2 Lifting winch. 3 Generator. 4 High tide level. 5 Low tide level. 6 Discharge.

Fig. 68 Disused tide mill at Trematon, Cornwall (see p. 198).

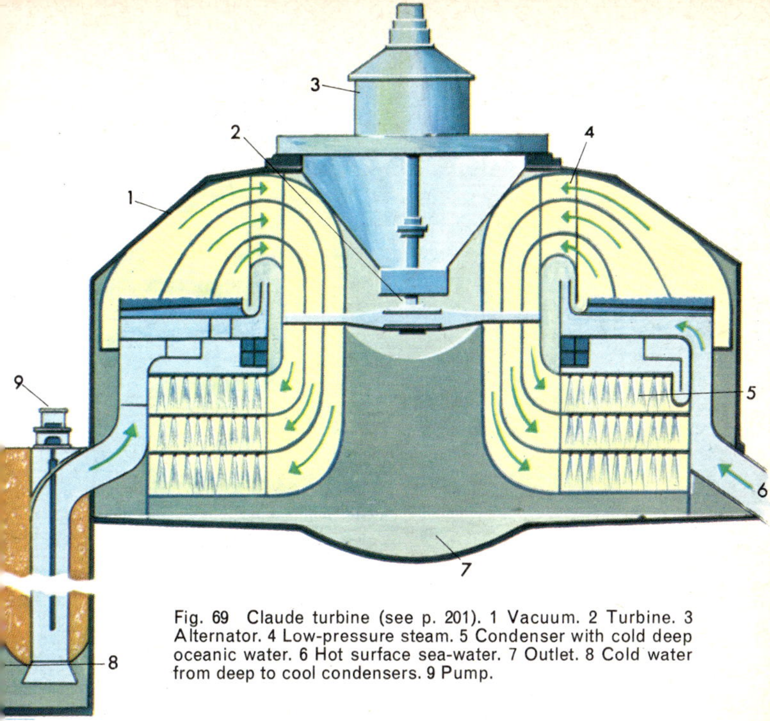

Fig. 69 Claude turbine (see p. 201). 1 Vacuum. 2 Turbine. 3 Alternator. 4 Low-pressure steam. 5 Condenser with cold deep oceanic water. 6 Hot surface sea-water. 7 Outlet. 8 Cold water from deep to cool condensers. 9 Pump.

Fig. 70 Possible use of cold deep water in power stations on suitable coasts (see p. 201).

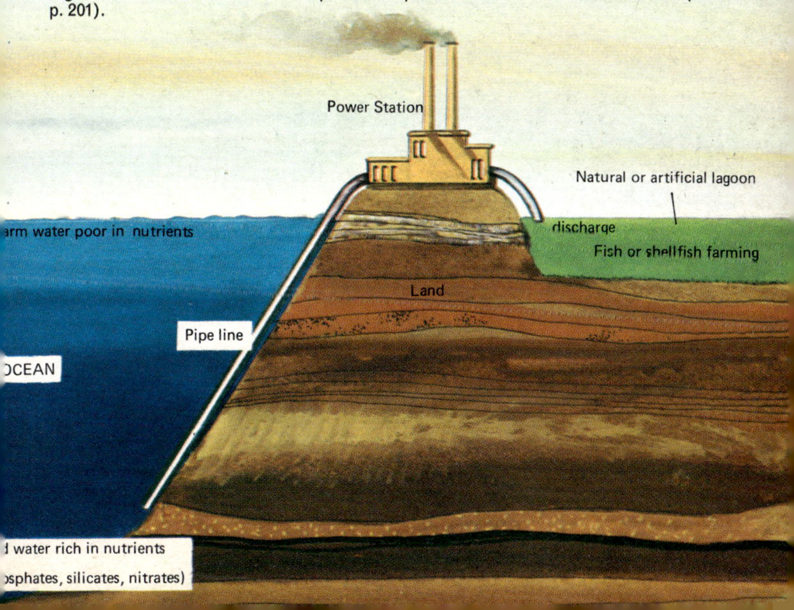

Power Station

Natural or artificial lagoon

arm water poor in nutrients

discharge

Fish or shellfish farming

Land

Pipe line

OCEAN

d water rich in nutrients

osphates, silicates, nitrates)

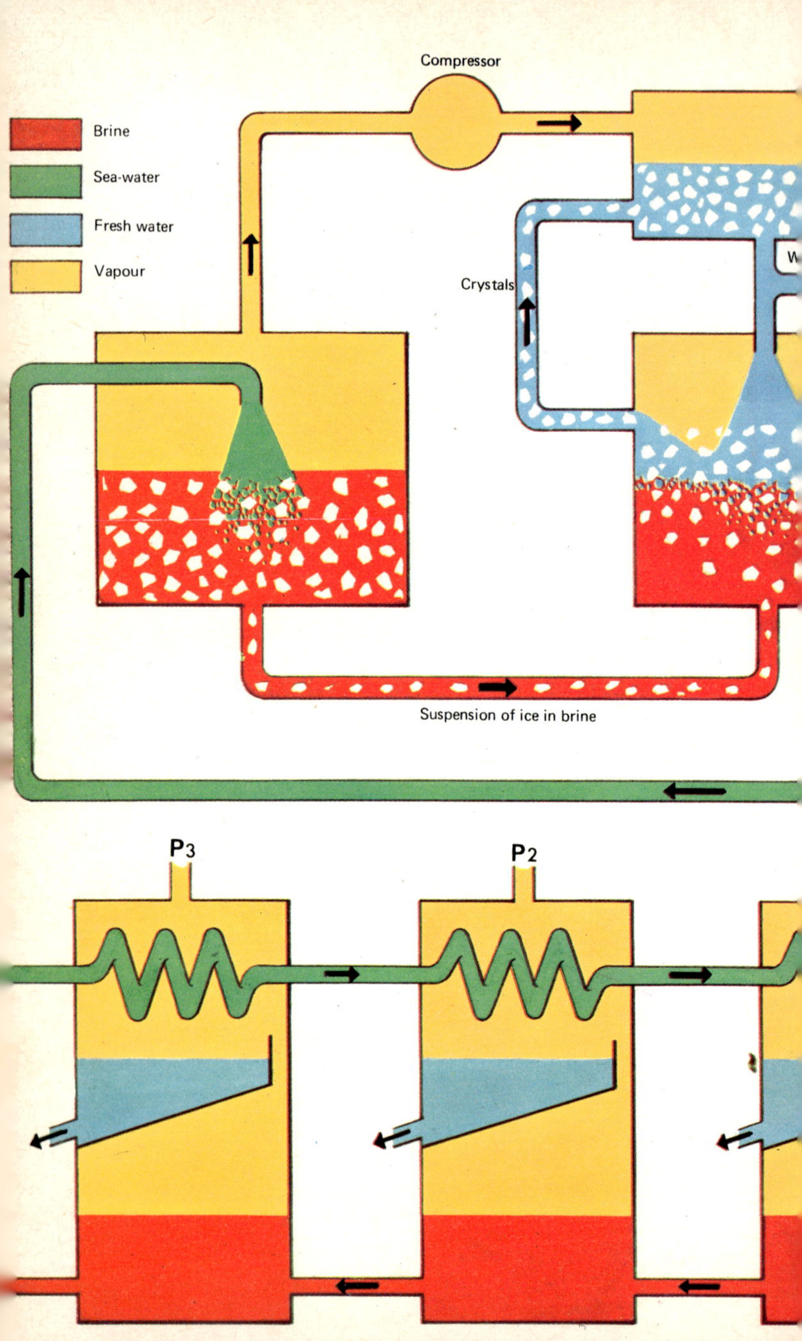
Compressor
Brine
Sea-water
Fresh water
Vapour
Crystals
Suspension of ice in brine
P3
P2

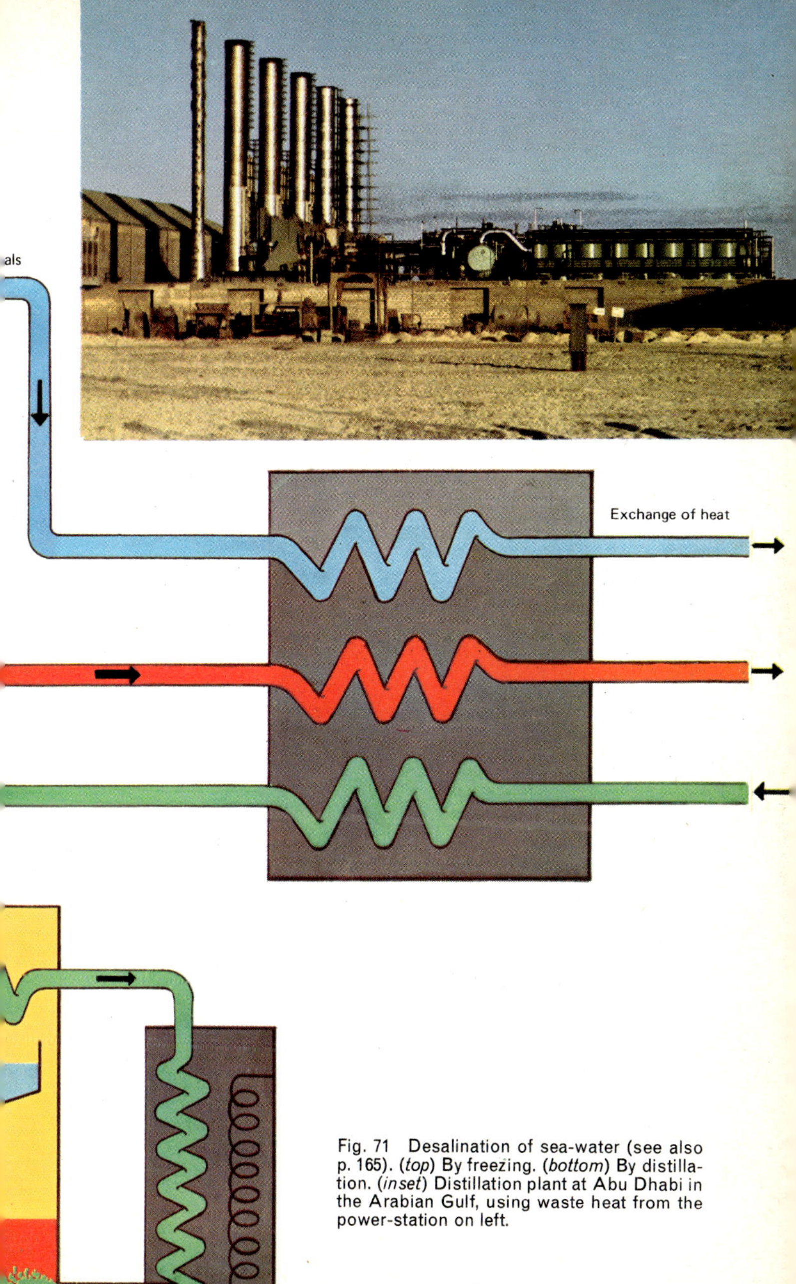

Fig. 71 Desalination of sea-water (see also p. 165). (*top*) By freezing. (*bottom*) By distillation. (*inset*) Distillation plant at Abu Dhabi in the Arabian Gulf, using waste heat from the power-station on left.

Fig. 72 Tanker aground, leaking oil which is being washed ashore.

Fig. 73 A seabird killed by oil.

Fig. 74 Spraying treated sand on an oil slick to sink it (the Shell 'Sand-Sink' method).

Fig. 75 *Torrey Canyon* oil fills Porthleven Harbour in Cornwall; it has been heavily sprayed with detergent (white slicks).

Fig. 76 A chiton (about 6 centimetres long) which has rasped clean an area of oiled rock. (Bahamas.) (see p. 237.)

Fig. 77 Some burrowing crabs have survived crude oil seepage into the mud and are throwing up both clean and dark oily pellets. (Saudi Arabia.) (see p. 237.)

Fig. 78 The construction of new oil refineries need not mean pollution and the end of local fishing. Traditional stake nets are still utilized effectively within the shadow of the Esso refinery at Slagenstangen, Norway.

Fig. 79 Effect of DDT pollution (see p. 236): thin-shelled pelican's egg, California.

Fig. 80 Stern-trawler fishing. (*insets, top right to bottom left*) Side trawler; tropical trawler catch; modern Dutch beam-trawler; handling the trawl in a rough sea; a big catch on deck.

SM 421

Fig. 81 (*left*) Drift netting. (*right*) Scottish general purpose diesel fishing craft for drift nets, purse-seines, ring nets and Danish seines, and capable of adaptation to inshore stern trawling.

(*left*) Ring netting (purse seining is on the same principle—that of surrounding a school of fish). (*right*) Purse seining; the fish are being scooped out of the full net with a lift-net.

(*left*) Long-lining as done by inshore boats. Deep-sea long-lines are laid on the seabed and may be miles long. (*inset*) Japanese floating long-line. (*right*) A sprat fishing boat deep in the water with a big catch.

Fig. 82 Danish seining. (*insets, left to right*) Hauling by seine winch and automatic coiler. The catch alongside.

Fig. 83 Pole-fishing for tuna.

Fig. 84 Japanese long-lining for tuna. Yellowfin tuna and albacore coming aboard. (*below*) Gaffing a yellowfin tuna.

Fig. 85 Chamber of Death at apex of tuna trap (see p. 217). Fish being gaffed out.

Fig. 86 Rope culture of oysters, Japan.

Fig. 87 (*above*) Adult sturgeon being netted in River Volga for hatchery purposes. (*left*) Sturgeon fingerlings from hatchery.

Fig. 88 (*left*) Ripe female salmon being stripped of eggs in hatchery. (*below*) Salmon alevins with yolksac, from a hatchery.

Fig. 89 Fish farming. A haul in a brackish-water fishpond in the Philippines. Milkfish jumping. (*below*) Sorting prawns on a Japanese prawn farm.

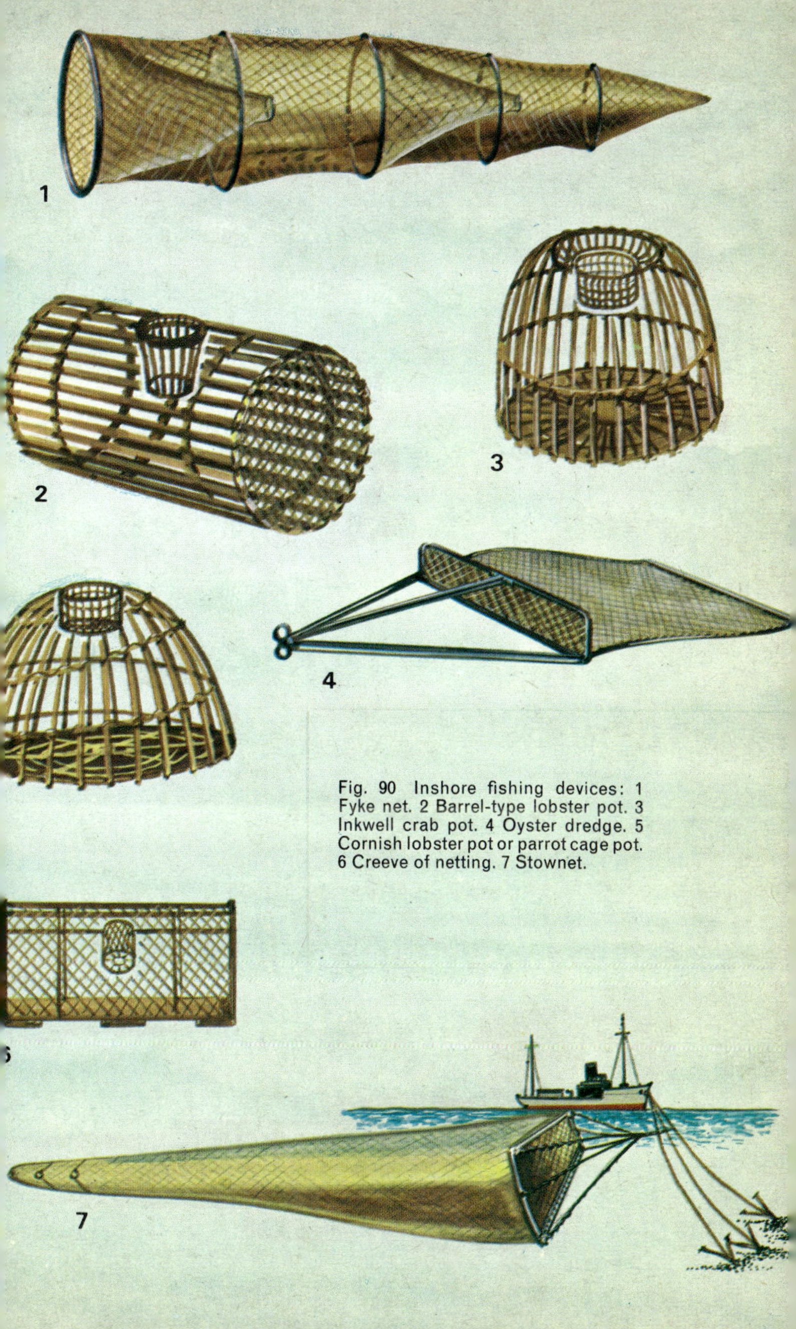

Fig. 90 Inshore fishing devices: 1 Fyke net. 2 Barrel-type lobster pot. 3 Inkwell crab pot. 4 Oyster dredge. 5 Cornish lobster pot or parrot cage pot. 6 Creeve of netting. 7 Stownet.

Fig. 91 Shellfish and crustacea. 1 Giant clam, *Tridacna*. 2 Queen. 3 Scallop. 4 and 5 Native oysters. 6 Queen. 7 Portuguese oyster. 8 Mussel. 9 Edible crab. 10 Prawn. 11 Lobster. 12 Crawfish.

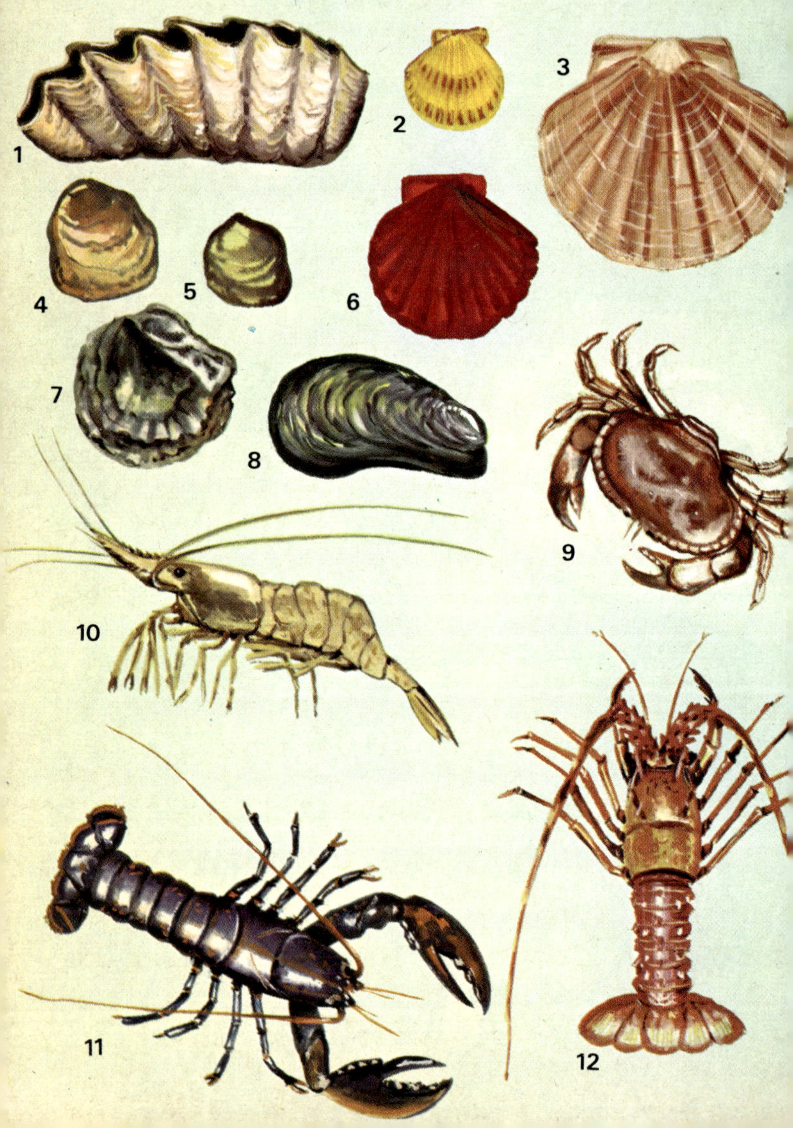

Figs 16–19 Some of the founders of modern knowledge about the seas and oceans.

Fig. 16 Commodore Matthew Fontaine Maury, U.S.N. (1806–73). In 1855 Maury published his classic book *Physical Geography of the Sea* in which was synthesized a monument of data concerning wind, waves and currents.

Fig. 17 Captain James Cook, R.N. (1728–79). Cook himself was principally a navigator and hydrographic surveyor, but his voyages also increased our scientific knowledge of the seas and oceans.

Fig. 18 Charles Darwin (1809–82). Darwin's round-the-world voyage (1831–6), as naturalist aboard H.M.S. *Beagle*, greatly contributed to his later discovery of natural selection, but in the field of oceanography it led directly to his theory of coral reef formation.

Fig. 19 Benjamin Franklin (1706–90). Franklin was a polymathic man with interests in many fields. In oceanography he is best remembered for his chart of the Gulf Stream which he published in 1770.

Fig. 34 The *Glomar Challenger.* During the latter half of the 1960s and the early 1970s, cores recovered during its numerous world-wide voyages have vastly increased, or radically changed, our previous knowledge about the past history of the Earth. The vessel employs highly-developed acoustical position-sensing equipment with automatic propulsion-unit control to provide dynamic positioning over a drilling station.

Forward of the 43-metre (142-foot) drilling derick (1) is the automatic pipe racker which can hold 7,000 metres (23,000 feet) of drill pipe. To position the vessel there are tunnel thrusters at the bow and at the stern (2 and 9) to provide for lateral/longitudinal manoeuvring. When the ship is on station, four hydrophones are extended below the hull to receive signals (3) from sonar beacons (4) positioned on the ocean floor. The signals are relayed into a computer which controls the thrusters to direct and maintain the ship's dynamic position over the drill hole. Re-entry of a drill cutting bit back into the drill hole is also fully automated using sonar.

A hole may be drilled 6·5 kilometres (4 miles) below the ship, and the drill head can penetrate through 750 metres (2,500 feet) of sediment and *in situ* bottom rock. Above the drill head are telescopic 'bumpers' (5) which absorb the variable motions due to waves. (6): the core sample tube (taking up to 10-metre (30-foot) lengths). (7): drill cutting bit. (8): the drill string which rotates the drill cutting bit for cutting and extracting the seabed cores (see also pp. 176–7).

Fig. 35 The food web, *key* 1 Albatross. 2 Tern. 3 Kittiwake. 4 Diving petrel. 5 Shearwater. 6 Cormorant. 7 Dolphin (whale). 8 Dolphin (fish). 9 Seal. 10 Flying fish. 11 Basking shark. 12 Baleen whale. 13 Herrings. 14 Medusae (zooplankton). 15 Krill (zooplankton). 16 Copepods (zooplankton). 17 Pteropods (zooplankton). 18 Crab larvae (zooplankton). 19 Salps (zooplankton). 20 Heteropods (zooplankton). 21 Ctenophores (zooplankton). 22 Chaetognaths (zooplankton). 23 Worms (zooplankton). 24 Diatoms (phytoplankton). 25 Flagellates (phytoplankton). 26 Seaweeds. 27 Squid. 28 Mackerel. 29 Bonito. 30 Tuna (tunny). 31 Swordfish. 32 Lantern fish. 33 Shark (toothed). 34 Sperm whale (toothed whale). 35 Large (giant) squid. 36 Hatchetfish. 37 Sablefish. 38 Octopus. 39 Anglerfish. 40 Viperfish. 41 Scarlet prawns. 42 Deep-water shark. 43 Gulper fish. 44/45 Swallower fish. 46 Bristle-jawed fish. 47 Anglerfish (with luminous lure). 48 Squid. 49 Glass sponges. 50 Crinoids. 51 Tripod fish.

Fig. 48 Seabed mining techniques. A. A crawler device – working below an ore carrier – using a rotary scoop to collect nodules. 1 Tube shaft for passenger lift to manned control module (4). 2 and 3 Tube shafts for material lift. 4 Manned control module where liquid and solid material is screened and crushed before being pumped to the surface. 5 Tube shaft for raw-material lift from the seabed. 6 Self-propelled crawler tractor with a rotary collector/scoop. B. An air-lift system mining device operated from a semi-submersible, self-propelled (propeller-driven) platform with a capability of operating in water depths of *c.* 4,000 metres (13,200 feet). 1 Towing cable. 2 Compressed air channel. 3 Mixer nozzles. 4 Hauling pipe. 5 Seabed collecting device. 6 Gliding skids. C. A schematic representation of part of a twin-sphere mineral lift device. With this system it is envisaged that the minerals on the seabed are brought to the surface by lowering a pair of water-filled spheres. One of the spheres is then emptied of water, and this causes minerals to be sucked into it. The water in the other sphere is replaced by air, and the combined unit then rises to the surface where the minerals are discharged into an ore-carrier vessel. The system is then recycled before its descent to the seabed. 1 Control valve. 2 Valve. 3 Upper sphere. 4 Seal. 5 Pump and air mechanism. 6 Lower sphere. 7 Discharge pipe. 8 Diaphragm (for creating suction to draw in seabed minerals). D. A manganese nodule miner bottom-crawler towed on a length of pipe from a surface ore-carrier vessel. The use of an articulated joint (4) allows the bottom crawler to remain firmly on the seabed and independent of the wave motion occurring to the surface vessel. 1 Ore storage tanks. 2 Mineral lift pipe. 3 Buoyant pump capsules. 4 Articulated joint (see above). 5 TV camera. 6 Light cluster. 7 Gathering head (incorporating scraper blades for scooping up minerals). 8 Sonar device.

Fig. 52 C.S. *Retriever, key*. 1 Bellmouth. 2 Officer's wardroom. 3 Radio office and chartroom. 4 Captain's quarters. 5 Wheelhouse. 6 Bow sheaves. 7 Auxiliary cable tension recorder. 8 Forward deep tank. 9 Cable machinery. 10 Hold. 11 Between decks hold. 12 Carpenter's shop. 13 Bo'sun's store. 14 Nos. 2 and 3 cable tanks. 15 Greasers' mess (seamen's mess, port side). 16 Engine room. 17 Greasers' quarters, starboard side; seamen's quarters, port side. 18 No. 3 cable tank. 19 Engineers' quarters (deck officers' quarters, port side).

Fig. 60 The food pyramid. Production of new material is by the plants in the bottom row, minute diatoms, peredinians and still more minute nanoplankton; and much less importantly by seaweeds. The latter break down to give detritus which nourishes many animals including the clams and lugworm. But the phyto-plankton feed the animals in the second row, and also the oysters, mussels and clams. These feed the animals in the third row, including the baleen whales, and the animals in the fourth row feed on those in the third row, including birds which give guano. Fishmeal is made from the small fish in the third row.

Fig. 71 Desalination. *Above*, by freezing: sea ice is fresh water, and sea-water can be artificially frozen in several ways to give ice crystals. In the *direct* method shown in the upper diagram, chilled seawater is sprayed into a vacuum vessel (left) where very rapid evaporation freezes a part of the sea-water. The resulting mixture of brine and ice crystals passes to a separating and washing tower, where the adhering brine is washed off the crystals with a stream of fresh water. The washed crystals are melted to give pure water. In the *indirect* freezing method, liquid butane, which boils at −10° C., is pumped into sea-water in a 'crystalliser' to freeze out ice crystals. Propane may be used instead of butane. *Below*, by distillation: hot, but not boiling, sea-water is run into a vessel at reduced pressure, where a part immediately evaporates and is condensed in the upper compartment of the vessel by cooling coils of incoming sea-water, which is thereby warmed. The residual brine passes onward through many similar stages, each at a successively lower temperature and pressure, and at each stage more water vapour is evaporated and condensed.

PART 2: MAN EXPLORES THE SEAS AND OCEANS

8 HISTORY OF MARINE EXPLORATION AND OCEANOGRAPHY

Man's interest in the seas and oceans has always been of the practical kind, and it still is. For early man, the sea was a source of food, and also, in common with other waters, the cheapest and easiest way of transport. A man can row or punt several tons, where he could carry only a hundred pounds or so.

At first, trips would be exploratory, venturing along the coast within sight of land. Thus the coasts of the Mediterranean and of western Europe became known; well before Christ Phoenecian traders visited Britain, and thus the frozen northern seas became known. Meanwhile, in the Pacific, the Polynesians were making long oceanic voyages, and in the Indian Ocean, Sindbad the Sailor and his fellow Arab seamen knew the secrets of the monsoons, namely the regular seasonal reversal in the direction of the trade winds, which blew them to Africa before the north-east monsoon in November to April, and to India and Ceylon before the south-west monsoon in May to September. These, and the world-wide fishing fraternity, must then, as now, have been keen observers of marine life; and we can be sure that much of the marine natural history known to Aristotle must have originated with talks to fishermen on sunny Mediterranean beaches.

Though, in the Middle Ages, most voyages were coastwise, the Vikings made their wonderful ocean voyages in the longships they had evolved, from Norway to Shetland, and to Iceland, Greenland and far down the east coast of North America. Their knowledge of what lay across the Atlantic may have accounted for the boldness with which Columbus sailed west four centuries later. Quite early in the Middle Ages, fishing voyages were made to the cod fisheries of Iceland from Hamburg, Holland and Britain. The now-extinct port of Dunwich in eastern England had a fleet of twenty Iceland cod-fishing boats between 1272 and 1307.

The Later Middle Ages

The Ottoman Turks overran the eastern Mediterranean in the fourteenth and fifteenth centuries, thus intercepting the rich trade in luxury goods which was carried to Europe from Egypt, to which it had come from the far east via the north-east monsoon and the camel caravans. The profits were incredible; and the Turkish conquests were the ruin of the Venetian merchant empire. There were the strongest motives for seeking an ocean route to the Orient which would by-pass the Turkish stranglehold. Byzantium fell to the Turks in 1453, and in 1492 Columbus crossed the Atlantic in an attempt to reach the Indies. In 1497, Vasco da Gama did succeed in reaching the Malabar coast of India, via the Cape of Good Hope; there he met the big four-masted junks from China, and the link was once more complete.

Exploitation of the east coast of North America was very rapid. The rich fishing grounds of the Newfoundland Banks were already well-known to fishermen even before Cabot made his voyage there in 1497. By the 1520s, French, Basque, English and Portuguese fishing vessels were working regularly off Newfoundland. Soundings were recorded along the coast westward from Cape Breton; the Gulf Stream was well known to British privateers in the sixteenth century, and its extension, the eastward-setting North Atlantic Drift, was observed in latitude 44°30′ North.

By Drake's time the Spanish and Portuguese colonizations in South and Central America, the Indies and Africa were of course well established, and private and official piracy and rivalry took British seamen to all these parts of the globe. Sir James Lancaster who, with Richard Hawkins, knew the secret of counteracting scurvy and berri-berri by the use of oranges and lemons, led the first British East India Company expedition which set out from Torbay in 1601, following his pioneer expedition in 1591. At the end of the seventeenth century William Dampier made an expedition to Australia in the *Roebuck*, which lasted two years. Dampier, who had already made a voyage round the world, had the curiosity and zeal of a born naturalist and was one of the first voyagers-cum-naturalist of the sort that was later typefied by Cook, Darwin and Moseley. It is not without reason that he has been called the Cook of a former age.

The Era of Captain Cook

The voyages of Captain Cook, one of the most famous of a group of seamen of several nationalities who largely completed the discovery of the Pacific Ocean, bring us to the beginning of oceanography. We may take his expert seamanship for granted; what may be less well known is that he was the first of the great explorers to use the marine chronometer. This notable advance in navigational accuracy was the result of a handsome prize of £20,000, a comfortable fortune in those days, offered by the English Admiralty for the best means of finding longitude. The prize was won by George Harrison, a carpenter's son, who built a watch compensated for changes of temperature and the movements of a ship at sea. Cook carried a copy of Harrison's chronometer No. 4, by Larcombe Kendall. Equally notably, at a time when scurvy so decimated the crews of ocean voyagers that ships were cast away because there were not enough fit men to handle the ship, Cook lost only three men out of 118 in a three-year voyage in unknown and dangerous seas, and none from scurvy. His successful anti-scorbutic was sauerkraut, a fermented and flavoured cabbage, and the compulsory ration was 900 grams (2 pounds) per man per week. Presented with the Copley Medal of the Royal Society, he said with typical modesty that he claimed no merit other than attention to his duty.

Joseph Banks, a well-to-do young man, sailed with Cook as naturalist, thus setting an example followed in many later expeditions. Banks made extensive collections of new animals and plants encountered on the voyage. Cook himself saw the halibut, salmon and sardine fisheries of the Red Indians on the north-west Pacific coast. He also took many soundings, some as deep as 1,350 metres (4,050 feet); and the Forsters, who sailed with him, made temperature observations and noted that the sea is stratified with depth. Cook was constantly logging the tides and currents and all sea phenomena, and he is remembered as the father of hydrography, particularly for his mapping of New Zealand, the South Seas and the north-west coast of America. Though he failed to make the north-east passage from the Pacific to the Atlantic, at the other end of the world he proved that, if a southern continent existed, it must lie south of 70° South latitude, and be icebound.

At about the same time Benjamin Franklin was making the first charts of the Gulf Stream, based on the collected observations of New England whaling captains, in pursuit of improved Atlantic crossing times for English mail ships (see p. 30).

The American naval officer Matthew Fontaine Maury, appointed officer-in-charge of the U.S. Navy's Depot of Charts and Instruments in 1842, is regarded by many as the real father of oceanography for his highly systematic collection of wind and current data from ships' log books over a period of ten years between 1840 and 1850, and for his book *Physical Geography of the Sea.* His were the forerunners of all modern charts which show current flow as well as indicating depth soundings.

Charles Darwin sailed with Captain Fitzroy on H.M.S. *Beagle* during her cruise of 1831–6. He was at that time a slim, tall young man, tough and athletic, able to rough it in company with Patagonian Gauchos, of whom he wrote 'they would cut your throat and make a bow at the same time', though he loved their company. Darwin never got over sea-sickness, but this distressing handicap did not prevent him from making many observations at sea; though of course his best-known work on this expedition was his study of coral reefs or atolls with its implications for the infant study of geomorphology and the evolution of the oceans (see pp. 75–6).

The 'Challenger' Era

Edward Forbes (1815–54), the Manx naturalist and geologist, who died at the early age of 39, was a pioneer of the study of the benthos, or bottom dwelling plants and animals of the sea. Using grabs and dredges, he showed that these creatures form communities, just as land animals do, and that, in similar environmental conditions, similar groups of marine animals occur. With the imperfect means at his disposal, Forbes was inclined to doubt whether life could exist in the deep sea below about 550 metres (1,800 feet). But already, unknown to him, Sir John Ross of the Royal Navy had brought up from over 1,800 metres (6,000 feet) in Baffin Bay mud samples containing many living worms, and also a living gorgon-headed starfish.

The laying of the transoceanic telegraph cables much increased practical interest in the depth and nature of the sea-bottom. Broken cables from great depths came up with living creatures attached. Wyville Thomson (1830–82), also a naturalist and geologist, Professor of Zoology at Belfast University in 1860, became interested in the marine benthos while looking over a collection made in deep water among the Lofoten Islands by the great Norwegian naturalist George Ossian Sars. With his friend, Dr W. B. Carpenter, Wyville Thomson organized expeditions to explore the deep waters of the continental slope along the north-west boundaries of Europe. Using the influence of the Royal Society, they were given the use of two small naval surveying vessels, H.M.S. *Porcupine* and *Lightning*. *Lightning* was described as a cranky little vessel, reputedly the oldest paddle steamer in the Royal Navy and scarcely seaworthy; but she did make soundings down to 1,100 metres (3,600 feet), and also obtained dredge samples from that depth. The *Porcupine* was a much better vessel, and aboard her the expedition was able to cover the continental slope from the Faroes to Gibraltar. Sixteen dredge hauls were made from more than 1,830 metres (6,000 feet), and two from over 3,650 metres (12,000 feet), and life was found at the deepest.

Wyville Thomson's name has been given to the submarine ridge which runs from Shetland to the Faroes at depths of about 550 metres (1,800 feet), and which separates the deep warm Atlantic water from the cold deep Arctic water to the east of the ridge. This contrast is striking in its effect on the abundance and nature of the fauna. The *Porcupine* is commemorated on the charts by the Porcupine Bank, a fine fishing ground far to the west of Ireland.

These successful exploratory voyages occupied the years 1868–70; in 1869 Wyville Thomson migrated to the chair of Natural History at Edinburgh University. The Royal Society then asked the Royal Navy to put up a seagoing vessel, and the money to run it, for a three-year voyage of exploration of the oceans of the world, taking in every aspect of physical, chemical and biological conditions. The House of Commons voted the funds 'with cordial assent', and now Wyville Thomson, appointed naturalist in charge, had to organize the great project, which ran from 1872 to 1876.

The ship provided was H.M.S. *Challenger*, a corvette of 2,306 tons, 76 metres (226 feet) long, and with auxiliary steam engines of

1,234 h.p. Above all, she had steam-winches to handle the great lengths of warp needed for sampling the deep water. Thus she was an adequate vessel for the long cruise, and the great amount of material she would collect; and the staff would have reasonable working and living space. But the organization was a severe test, for little was known about instrumentation, and most had to be invented or improvised. All honour, then, to the success of a voyage which founded a science, the new science of oceanography, and the results of which filled fifty bulky tomes. (One of its most significant achievements was the discovery in the course of soundings of all five oceans more extensive than had ever been carried out before of the Mid-Atlantic Ridge – the first intimation that the deep ocean bed was not simply a smooth plain of sediment.)

Meanwhile, America produced an oceanographer of the first rank in Alexander Agassiz. Trained as a mining engineer, and with a personal fortune derived from copper mines, he was able to finance and take part in research voyages, and to support the Harvard Museum of Comparative Zoology. His skill as an engineer enabled him to devise new intruments for deep sea research, and he was the first to use thin steel wire for this purpose, instead of the unhandy thick hemp lines and ropes of the earlier expeditions.

Among many later expeditions sent out to follow up and extend the work of the *Challenger* were those of another rich man able to follow his own bent, namely Prince Albert of Monaco. After serving in the Spanish navy, he became keen on oceanography and, first in the small *Hirondelle*, and later in the larger *Princesse Alice*, he made many research expeditions and pioneered the large-scale use of floats for plotting the speed and course of ocean currents. He also founded the famous Museum at Monaco, and an Oceanographic Institute in Paris. One of his interests was the deep midwater region of the ocean, especially since a wounded sperm whale, charging his yacht, disgorged the tentacles of a freshly caught squid of very large size, much larger than any so far known. Such big, strong and fast-swimming animals must clearly be very difficult to capture. But somewhere in the deep midwater there is a food-chain still not completely known, which leads up to these great squids, and thence to the sperm whales (see p. 213).

In 1872 the first shore marine biological station was opened in

Naples. This was followed in 1888 by the British Marine Biological Association's Plymouth laboratory and in 1892 by the Hopkins marine station of Stanford University at Pacific Grove, California. Prince Albert of Monaco set up his museum and institute in 1906. In 1902 was founded the Marine Biological Association of San Diego which, as the Scripps Institution of Oceanography, was to become one of the most important centres for oceanographic research in the twentieth century, together with the Woods Hole Oceanographic Institute, founded in 1930.

By the 1890s, the first seemingly boundless wealth of the commercial sea fisheries of northern Europe was in decline. Those who held that the riches of the sea were inexhaustible had to think again; and, in the alarm which followed, the maritime countries of north-western Europe came together to establish in 1902 the International Council for the Exploration of the Sea, with headquarters in Copenhagen. Oceanography was an integral part of its programme from the start, and it has since flourished and survived two world wars, doing excellent work. It has set the example for the establishment of similar regional bodies elsewhere.

In the years after the *Challenger* finished her work, other historic oceanographic vessels such as the U.S.S. *Albatross*, the U.S. survey steamer *Blake*, the French *Travailleur* and *Talisman*, the Russian *Vitiaz* and the German *Valdivia* were ploughing the seas.

More Recent Times

In 1910 Sir John Murray, one of the *Challenger*'s original staff, put up the money for an Atlantic cruise by the Norwegian research vessel *Michael Sars*, with Johan Hjort, the great Norwegian fisheries expert, in charge. The result of this well-organized little expedition by two distinguished oceanographers, who were also close personal friends, was the book *The Depths of the Ocean*, which is a classic. Beautifully illustrated, it blends the results of the expedition with previous knowledge, in a series of chapters by specialists, and with so light a touch that interest never flags.

Before this, the great Norwegian polar explorer and oceanographer Fridtjof Nansen had organised the 1893–6 *Fram* expedition, in which he drifted for three years in the Arctic pack-ice. In his subsequent career as Professor of Oceanography at Christiana

University, he made very considerable advances in the theory of the interrelation between winds and currents and added pieces to the ocean circulation jigsaw with his studies of the Arctic Ocean basin. He also made important contributions to the development of instrumentation, such as the Nansen–Pettersson bottle for measuring temperature and salinity at differing depths. By its use fundamental information for the study of surface and deep-water currents is obtained (see Fig. 59).

World War I put a stop to most work on the oceans. Denmark, however, was a neutral in that war, and continued to run a steamer service to her then West Indian colonies. From these ships, and from research vessels after the war, the great Danish biologist Johannes Schmidt organized the collection of plankton samples in his search for ever younger stages of the European eel. Thus he was able gradually to narrow the probable area of breeding, until he found the eggs and newly hatched young in the ocean south of Bermuda. 'Here is the breeding ground of the eel' he was able to write proudly in 1920.

The Danish ship *Dana* explored the Indian Ocean between 1920 and 1922, during which time was discovered the underwater rise subsequently named the 'Carlsberg Ridge' after the famous Danish lager company who contributed funds to the expedition.

In 1926, Germany re-entered the oceanographic field with the expedition of the gunboat *Meteor* to the south Atlantic. This cruise was ostensibly to search for gold in seawater, in the hope of paying off the preposterous load of reparations, payable in gold, resulting from the peace treaties of 1919. It is difficult to believe that Professor Fritz Haber, who must have known that gold occurs in sea-water as the trace of a trace, really expected success; but, however that may be, the *Meteor* expedition was an abiding credit to the resurgent German marine science. Its results, worked up in conjunction with the results of the British *Discovery* expeditions, helped to apply the lessons of meteorology to the oceans: the 'fronts' of the atmosphere became the 'convergences' of the hydrosphere and great practical and theoretical advances were made. The *Meteor* expedition is also remembered as being the first to make extensive use of echo sounding, then in an emergent state and perfected during World War II. Seventy thousand soundings were made, compared with the 2,000 which was the most achieved by any previous expedition, and for the first time the extreme ruggedness of the ocean bottom became apparent.

In 1929 the American vessel *Carnegie* explored the Albatross Rise in the Pacific – first discovered by Louis Agassiz – and revealed an oceanic mountain range.

The series of commissions of the *Discovery* and the *Discovery II*, throughout the inter-war years, were organized to follow the fate of the whales of the Southern Ocean, their food and feeding, migrations, breeding, growth and rate of mortality. It was a comprehensive study of the whole of the environment of the whales. It was no fault of these devoted men that the march of the whale stocks towards extinction was not halted. The results were published in a splendid series of reports on the distribution of the whales, their food-chain and the hydrography of the ocean, with accounts of forays up the coast of South America to investigate the Humboldt Current (Peru Current), to the Falkland Islands for a trawling survey, to South Africa and on into the Indian and Pacific Oceans. The great Southern Ocean became the best-known of all the oceans.

In the years following World War II the great traditions of oceanographic exploration continued as before. Hans Pettersson's *Albatross* Expedition of 1947–8 added to knowledge on currents, marine life and bottom profiles in the tropic zones. It also made revolutionary advances in bottom coring techniques and pioneered the use of seismic echo methods for finding seabed sediment thicknesses.

In 1950–1, during its world cruise, the hydrographic survey ship H.M.S. *Challenger* explored the foundation rocks of several Pacific coral reefs. The Danish *Galathea* Expedition of 1950–2 found marine organisms 9 kilometres (6 miles) down at the bottom of the Philippine Trench.

C.F.H.

The 'Glomar Challenger' Era

The most recent advances in marine exploration and oceanography stem from about the time of the International Geophysical Year 1957–8, when the scientists of all nations and disciplines joined forces to make a combined attack on some of the outstanding problems concerning the nature of the Earth. Then in 1958 a co-operative, full-scale investigation of a single ocean was proposed by the International Council of Scientific Unions, and the Indian Ocean was chosen. More than twenty-five nations sent either ships or scientists, and data centres were established in the U.S.A., the Soviet Union and India. This expedition officially began in 1959 and continued right through until as late as 1965. Many new oceanographic features were discovered and explored, and subsequently many new ideas evolved concerning the problems about the nature and evolution of the great ocean basins.

By the early 1960s possible new commercial exploitation of the oceans was also a major factor in promoting new research programmes. In addition the problems of men living in outer space triggered several experimental, manned seabed habitat systems in which the U.S. astronauts took part to explore the physiological and psychological problems of living for long periods in unfamiliar stress-inducing environments. The strategic/economic problems of the oceans also played a large part in moulding the nature of the newer oceanographic research. The new, unrealized economic possibilities of ocean exploitation led several governments to subscribe large annual budgets to further this research. Since the lull following the first excursions into cosmic space, even more national finance is now being channelled into hydrospace where economic and strategic rewards appear more readily attainable.

From the time of the advent of the International Geophysical Year, hundreds of vessels have taken part in various aspects of oceanography and maritime studies and it would be invidious to name only a selection. In the United States alone more than ninety such vessels are involved in various programmes.

However, one vessel must be singled out for the unique role it has played and is still playing in marine science. When the new U.S. drill research ship *Glomar Challenger* (see Fig. 34) entered service in 1968, it truly heralded a new age in oceanography as the first *Challenger* did in 1872. Through the wide-ranging voyages of *Glomar Challenger* the oceanographers have realized their dream of being able to drill holes into the deep ocean floor and discover a wealth of data about the past history of the oceans and the Earth itself. Ocean-floor spreading was accepted by sceptical scientists only after studies of the long unique cores drilled from the seabed by *Glomar Challenger*. In the century since the circumnavigating voyage of the *Challenger*, the single most important oceanographic discovery has been that of ocean-floor spreading.

Nevertheless, we must not lose sight of the fact that the aims of the newer marine exploration and oceanography is for man to get to know the seas and oceans in all their aspects. It concerns man's use of the sea for multifarious and diverse purposes such as recreational sites, mariculture, mineral extractions, seabed engineering projects and waste disposal.

In many areas of the world fishermen still employ fishing techniques that were for the most part used in biblical times. Nowadays much of the effort of conducting commercial fishing is expended in the search for schools of fish or sea creatures which are large enough to warrant exploitation (see p. 229). The world's fish catch has tripled in the past thirty years. Of the 50 million metric tons of fish caught annually in the later 1960s, only one half was consumed directly by man as food, and the remainder was used much less efficiently to feed animals (see pp. 229 and 234).

The economic pressures for exploiting the resources of the sea are wide-reaching. Apart from utilizing marine creatures as a source of food, minerals are also of prime importance in the future world economy. The demand for hard minerals such as gold, copper, chromium, nickel, etc., will, by 1985, have doubled that of the 1970s, and by the year 2000 the demand will have tripled. By the end of the present century the demand for fossil hydrocarbons will be *at least* four to six times what it is in the 1970s. The total volume of potential oil- and gas-bearing rocks is clearly limited. On land most of the large oil fields have now been found. To meet the ever-growing demand, the resources of the continental shelf must necessarily be exploited. It is not without reason that in the 1970s the North Sea has been aptly nicknamed 'Europe's newest frontier'.

The science of the seas and oceans has undergone many changes since Victorian times. For example, at the time of the *Challenger* expedition, marine biologists were solely concerned with collecting and identifying forms of marine life. A century later, their research objectives are directed to developing a thorough understanding of physiological phenomena, ecological responses and behavioural processes.

The new generation ice breakers, hydrofoils, hovercraft, deep-diving submersibles, drill ships, underwater habitats and factory ships have all played their role in developing modern oceanography. Many of the great aircraft manufacturers have diversified into ocean-based projects such as rescue submersibles, devices for clearing up oil spillage and seabed petroleum recovery systems. Nowadays oceanographic expeditions have computers, navigational systems linked to earth-circling satellites, sonar and underwater television. Many research ships are floating laboratories utilizing analytical techniques like X-ray diffraction, thermal-neutron activation, atomic absorption, C-14 dating and mass spectrography. Modern oceanography is truly the amalgam of many sciences and diverse disciplines, and aboard research ships today, biologists, atomic scientists, micropaleontologists and highly skilled drilling crews work shoulder to shoulder in wresting the secrets from beneath the waves.

P.L.B.

9 TECHNOLOGY AND THE OCEANS

Although oceanographic technology has antecedents stretching far back in history – for example, Alexander the Great experimented with a diving bell, and in the seventeenth century diving techniques were surprisingly advanced – the revolution in sub-surface technology did not come about until after World War II. The Phoenician frogmen who destroyed the city of Tyre in 332 B.C. must have had some form of diving equipment, but little is known about its technical specifications. All modern diving equipment stems from the invention of an Englishman, Augustus Siebe, who in 1819 designed the now familiar watertight suit with a helmet and glass viewing window.

In 1930 the first of the modern bathyspheres was invented by Otis Barton, an American, and in 1934 Barton, accompanied by another American, William Beebe, descended more than half a mile below the surface. But a number of different factors triggered off the modern era in underwater technology and exploration. Among these were Cousteau's invention of the aqualung; the space programme; economic pressures to find new sources of raw material and food; and two significant disasters. It was probably the two disasters which provided the most important impetus of all.

The first of the disasters was the loss of the U.S. atomic-powered submarine *Thresher* which sank off Cape Cod in over 2,400 metres (8,000 feet) of water on 10 April 1963. At this time no submarine rescue vessel could descend lower than 450 metres (1,500 feet), yet this was two years after Piccard and Lt Walsh had succeeded in penetrating to the bottom of the 11,040-metre (36,000-foot) Marianas Trench off Japan. The second disaster was the loss of a U.S. hydrogen bomb off Palomares in 1966.

Within a year of the *Thresher* disaster the United States Navy began its Deep Submergence System Project (DSSP), which was the first major undertaking of its kind in manned ocean engineering. Its primary task was to construct six submersible vessels capable of reaching a stricken submarine anywhere in the world and mounting a rescue operation to depths up to 1,500 metres (5,000 feet) of water. It was from this date that significant all-round technological progress in hydrospace began.

There are numerous parallels between the problems of interplanetary space and hydrospace. In both environments humans are exposed to hostile conditions, and fail-safe life support systems need to be evolved for man to survive in them. Since World War II, methods of oceanography have undergone a great transformation. Previously oceanographic research, for the most part, was conducted from the decks of small vessels by suspending trawl nets below the surface in what amounted to simply a blind groping beneath the surface, grabbing samples willy-nilly. Jacques Piccard succinctly remarked it was like 'a blind man making a butterfly collection'.

Piccard himself was one of the great pioneers in bringing about a change of method when, in 1948, he launched one of the pioneer submersible vessels, the bathyscape *FNRS II*.* Later he built the famous *Trieste*, which in 1958 was purchased by the United States

* *FNRS I* was Piccard's aeronautical balloon.

Navy and then modified to make the famous Marianas Trench descent in January 1960. Another pioneer vessel, the diving saucer *Denise*, was designed and built in 1959 by Jacques Yves Cousteau, and this descended to about 300 metres (1,000 feet).

In the 1970s, hundreds of submersible vessels are operational. Some have been built for pure oceanographic research, but many are designed to operate in commercial mining and drilling operations on the seabed. Submersibles differ from classic military submarines in that they are generally able to withstand greater pressures and are more manoeuvrable. The main purpose of submarines is to submerge just deep enough to escape detection from the surface. Very few submarines can go deeper than 300 metres (1,000 feet). Submersibles on the other hand are designed to operate at any depth, as deep as 11,040 metres (36,000 feet) if so required. Many of the submersibles can operate independently like submarines, but some are designed to operate from mother vessels or are designed to work as robot vessels operated from the surface. The range of submersibles extends to many different forms. There are mobile television platforms, bulldozer machines for bottom mining and the Soviet Union's 'glider' submersibles for towing at about 90 metres (300 feet) below the surface to monitor fishing with trawl nets (see also Fig. 41).

In addition to the submersible vessels, some remarkable surface vessels have been developed such as drilling vessels like the 120-metre (400-foot) *Glomar Challenger* (Fig. 34), which can drill holes in the seabed in over 6,000 metres (20,000 feet) of water without traditional anchorage methods. Positioning is maintained over the drill hole by a computerized system of pulses from acoustic beacons set on the sea floor, and these are then picked up by a ship's mounted hydrophone array and processed. Corrective action is fully automated, and if the ship begins to drift off target, the necessary corrective action is made by special propulsion units.

Another remarkable surface vessel designed for oceanographic research is a 106-metre (355-foot) long floating instrument platform named Flip, which has been developed by the Scripps Institution of Oceanography, California (Fig. 44). This remarkable vessel can be towed to the investigation site; then, by flooding its ballast tanks, made to 'flip' into a vertical position and rest on the bottom.

Some of the most remarkable of the modern surface vessels are the various types of drilling platform designed for oil and gas exploitation on the continental shelves (see 'Oil and Gas' below).

Undersea Life Support Systems

Hand in hand with the development of submersible vessels has gone the research and development of support systems to keep man alive under the oceans. Already submersibles have been found to be more reliable life support systems than traditional diving suits. In North Sea exploration alone, four divers were killed in one year's operations.

The major difficulty in man's free submergence to great depth arises from breathing the nitrogen in air at high pressure. Because nitrogen diffuses slowly through the lungs, it is retained for a long time in the body. When pressure is reduced during the diver's return to the surface, the nitrogen is released as bubbles in the bloodstream resulting in decompression sickness known as the 'bends'. The bubbles block the blood circulation and so deprive nervous tissue of oxygen, producing severe pain and causing paralysis or even death. Nitrogen at high pressure also causes nitrogen narcosis – a kind of intoxication – possibly due to nitrate or hydrazide formation in certain cells in the body.

By allowing a diver to rise slowly to permit diffusion to occur, without inducing bubble formation, it is possible to avoid the bends, but the time involved is often considerable. For example, for a 150-metre (500-foot) dive, decompression may last as long as three days. The ideal situation is that man should ascend as quickly as he descends, and underwater life support system technology is chiefly concerned in bringing about this ideal.

In experiments, helium has been tried as a replacement to the troublesome nitrogen, but its use has its own problems such as availability throughout the world and the difficulty in regulating its flow due to its lightness. It has also been shown to produce physiological side effects, chief of which is reducing the ability of the human body to endure long periods in cold water. Helium also reduces human conversation to an unintelligible gabble, so that traditional voice communication is rendered impossible. Although helium certainly reduces the narcotic effects of high pressure nitrogen, it is likely that helium itself may produce similar effects if used at great depths.

The drug PDHA* (which needs injection) has proved effective in experiments involving rats. The drug lowers the level of fats in the blood stream, and it has been known for some time that fat people are more susceptible to bends than thin people are. One theory is that the PDHA demolishes the veneer coating of fat that builds up and protects each nitrogen bubble in the blood stream.

The question has often been asked as to how whales are able to dive to great depths and quickly surface again without apparent ill effects (see also p. 68). However, it appears that through the long process of evolution the whale has developed a mechanism which has enabled it to survive. Even if we understood it, it is doubtful that man could adapt his own physiological system to cope with a rapid change in environment. Another line of research has involved artificial gills to tap oxygen directly from the water at depths, as fish do, but this seems an unpromising line to pursue, because sea-water contains insufficient dissolved oxygen to support metabolism in mammals.

When a diver's tissues have become saturated with the high pressure gas mixture after a few minutes, additional time spent down below will not increase tissue saturation nor decompression time. To make maximum use of submergence time, it becomes economic for a diver to spend long periods under water rather than spend time in wasteful decompression schedules.

In an early experiment two French divers remained one week below the surface in the Mediterranean in Cousteau's first underwater house *Conshelf I*. According to the divers' comments their main difficulties were psychological and not physiological as had been expected. In further experiments with *Conshelf II* in the Red Sea, and with *Conshelf III* in the Mediterranean, Cousteau's team has gained great familiarity with the problem of prolonged living on the sea bottom.

In the United States, the *Sealab* experiments were undertaken shortly after the *Thresher* was lost. As a consequence of this work, and work done by the Soviet Union in the Black Sea, it is now

* Partially depolymerized hyaluronic acid.

known that man can survive for long periods on the sea bottom at considerable depths if the breathing mixture is stringently maintained at correct levels.

The Mineral and Chemical Harvest

It has long been known that in every 100 kilograms (219 pounds) of sea-water there are 3·5 kilograms (7·6 pounds) or $3\frac{1}{2}$ per cent of mineral matter (see also p. 8). The percentage would be higher but for the fact that the two commonest minerals, lime and silica, are removed from the sea by shell-forming animals, and the minerals that remain, with a few exceptions, are not extracted by animals of the sea.

Sea salt was the first mineral that ancient man succeeded in mining from the oceans with his primitive technology. Three-quarters of 'sea salt' is common sodium chloride (domestic table salt) and can be obtained from the sea by the simple expedient of allowing sea-water to evaporate naturally in shallow solar ponds and then harvesting the residue. In principle, modern methods differ little from those practised by ancient man. Salt is 'mined' from the sea all over the world. In Australia a vast new industry has been created in those parts which experience high evaporation rates such as Shark Bay, Port Hedland and South Australia (Fig. 47). In 1972 more than 4 million tons of sea salt was exported to Japan from the West Australian solar salt-pans alone. Magnesium has been extracted from sea-water for many years and as we have seen over 65 per cent of the world's annual consumption of magnesium is at present won from the sea. In every cubic mile of sea-water there are 4–6 million tons of magnesium, and much of it can be extracted by using modern chemical techniques (see Fig. 47).

Bromine is another element which is currently extracted from sea-water in large quantities, and in every cubic mile of sea-water there is a quarter million tons of it. The method of extraction was invented in 1933, and bromine was the first mineral obtained from the sea using chemical technology. It became an important mineral in the 1930s owing to the large demand created by the expanding photographic industry. In addition, the oil companies discovered its use in the manufacture of anti-knock petroleum compounds.

Gold from the sea has long fascinated mankind. In every cubic mile of sea-water there are about 17 kilograms (38 pounds) of gold in fine suspension. Although this sounds like a great deal of material, it represents very little in relation to the bulk of sea-water. The German *Meteor* Oceanographic Expedition of 1924 was supposedly instituted with the idea of commercially extracting gold from the sea in order to pay Germany's World War I reparation debts. However, it seems unlikely that the scientists of the expedition truly believed that this was economically practical. It is more likely that they saw it as a first class opportunity of mounting an expensive oceanographic expedition, and once they had put to sea, most of the work of the expedition was biological or oceanographic in content. Nevertheless, gold holds a bright promise in future seabed-mining operations. Disregarding the suspended gold in sea-water, vast quantities of alluvial gold are contained in offshore sedimentary deposits in Alaska, Australia and New Zealand. These deposits will be exploited in the future by deep-sea dredging, using methods not unlike traditional gold-dredging operations in the rivers of New Guinea and Alaska.

Iodine is one of the scarcest elements in the oceans. However, the oceans were until recently a principal source of iodine, for seaweeds concentrate iodine from sea-water, and its extraction from seaweed is comparatively easy. In the free ocean there is only about one gram of iodine in 80 tons of water. In seaweed it becomes concentrated in the ratio of one gram of iodine to 200 grams of dried weed. Seaweeds are not the only marine organisms capable of concentrating important elements or minerals. Cobalt is concentrated by lobsters and crayfish, and radiolarians concentrate appreciable amounts of vanadium, while oysters are able to concentrate copper.

Manganese is one of the most important minerals on the actual seabed. The *Challenger* Expedition was the first to dredge up a manganese nodule in 1872, but it is only in more recent times that vast areas of deep ocean floor have been located and surveyed and found to be strewn with dense concentrations of manganese nodules. In addition, these nodules have been found to contain other important economic concentrations of minerals such as nickel and copper.

The formation of manganese nodules on the seabed has been a great puzzle to mineralogists. It seems that a nucleus is first required in order that the nodule formation can begin. Nodules which have been sectioned for examination show a variety of nuclei; some contain several kinds of nucleus material within the same nodule, such as fragments of volcanic rock, shark teeth and whale earbones. The nodules vary in form and size. They are mostly about 2 centimetres ($\frac{3}{4}$ in.) in diameter, but they may vary from very tiny objects to lumps almost 30 centimetres (12 inches) in diameter.

Each area of the seabed has its own characteristic nodule chemistry, and the geochemical content is determined by the environment where growth takes place. Thus a 'manganese' nodule in the laboratory can be indentified with a specific locality by its tell-tale concentration of different metals.

Manganese and other metals are continuously being supplied to the sea by rivers and by the leaching action of the sea itself. Nickel and copper are much more abundant constituent minerals in deep-water nodules recovered from 4,000 to 6,000-metre (13,000 to 20,000-foot) depths. Cobalt and lead are more abundant in shallow-water nodules, in particular in those that occur in less than 1,000-metre (3,300-foot) depths.

About 50 per cent of the volume of a nodule consists of open pores, and most of the metal oxide content is manganese and iron. Nickel constitutes on average from 0·2 to 1·8 per cent, copper 0·1 to 1·6 per cent and cobalt 0·1 to 1 per cent. In addition there are lesser amounts of vanadium and molybdenum.

The problem which has been confronting technicians for several decades is how to harvest these deep-water nodules. However, the problem remained an academic one until world demand for strategic minerals alerted authority that sea-floor resources might prove more important in the long term than resources ashore.

Recently methods have been developed to suck the nodules to the surface using a technique of liquid-air (hydro-pneumatic) lift. The nodules are first collected by dragging a large-sized scoop along the bottom which is towed from either a traditional vessel or a self-propelled semi-submersible platform similar to modern drilling rigs (see Fig. 48). After collection the nodules are transferred to ore-carriers which ply between the shore-based smelting/separation plants and the deep-sea mining fields.

Another suggested method of lifting nodules is called 'Twin-sphere Mineral Lift' (Fig. 48). In this method the minerals (nodules) are brought to the surface by lowering a pair of water-filled spheres to the seabed. One of the spheres (the bottom one) is emptied of water, and this causes minerals to be sucked into it. The water in the other sphere is then replaced by air to supply buoyancy, and the combined unit then rises to the surface. At the surface, water is again pumped into the lower sphere, and the minerals it contains are forced out through a discharge pipe. The cycle can then be repeated by reflooding the upper sphere.

An alternative idea, not yet fully developed, is to leach out the copper and nickel *in situ* by using dilute sulphuric acid and concentrated ammonia. It has been estimated that by using leaching solutions, up to 95 per cent of the metal could be extracted in a period of about six hours. Since the nodules are highly porous, it should be possible to extract most of the copper and nickel content without crushing or disturbing the nodules. In the method envisaged, the process would involve piping the solvent liquids to the sea bottom and releasing it directly over the nodule beds. The 'pregnant' saturated liquid would then be collected downstream and pumped back up to the surface. In such a way the beds would be conserved and later utilized again when the nodules would have had opportunity to concentrate more metallic salts.

It has been estimated that half the free world's demand for cobalt could be gleaned from 260 square kilometres (100 square miles) of seabed when mining operations begin on a commercial scale.

The waters and beds of the world's seas and oceans are rich in many important economic minerals. For example, uranium can now be extracted from sea-water, but the process is not yet commercially practical. In the Gulf of Mexico 3,000,000 tons of sulphur are flushed up annually. Along the coasts of Britain, Malaysia and Thailand large quantities of alluvial tin lie offshore. In many parts of the oceans, nodules of phosphate rock – important in fertilizing products – await future exploitation.

One of the most promising areas for seabed exploitation lies on the bed of the Red Sea where a 90-metre (300-foot) thick deposit, rich in copper, zinc, lead, silver and gold, has been proved. One of the great obstacles to mining in the future will not be with mining techniques or extraction methods. Based on past experience we can be assured that man's ingenuity will solve any problem in the sphere of technological 'know-how'. The main problem will be concerned with territorial rights and ownership of the seabed containing the deposits. For example, the mineral deposits in the Red Sea are claimed by both the Sudan and Saudi Arabia. Another problem involves the economics of deep-sea mining. It has been pointed out that unrestricted ocean mining by international groups, able to capitalize initial development, could involve small land-locked countries, which at present survive by their mineral exports, in economic ruin. No one can yet foresee in detail the implications of what has been called the menace of an uncontrolled Klondyke over five-sevenths of the Earth's surface. (See also 'Who Owns the Seabed?' pp. 237–9.)

Oil and Gas

The oil and gas industries were the pioneer industries in exploiting the mineral resources that lay beneath the seas and oceans. Commercial oil prospecting on the sea floor began off the coasts of Southern California in 1891 from piers abutting into shallow water, although widespread activity in the sea did not begin until the 1930s when operations began in the Louisiana bayous. By the 1940s, wells were being drilled in the shallow waters of the Gulf of Mexico, but international offshore exploration did not get under way until after World War II.

There is now world-wide activity on many of the great continental shelf structures where oil and gas are trapped in hardened sediments that were deposited in prehistoric seas.* The continental shelves are estimated to contain up to 50 per cent or more of the Earth's oil and gas resources. To exploit fully these vast reserves many technological innovations have come about, and the technological demands of the oil companies have greatly stimulated invention and development (see p. 191).

At present there are three main methods of drilling the seabed and maintaining wells.

1 From a platform (fixed, jack-up or floating) or a ship.
2 From remote-controlled equipment on the seabed.
3 From manual equipment on the seabed fixed at atmospheric

* For example, there are over 8,000 wells on the U.S. continental shelf alone.

metres
SEA LEVEL
million years
Sand
QUATERNARY
1·85
300
Clay
TERTIARY
70
600
CHALK
Limestone
CRETACEOUS
Clay
135
JURASSIC
1,200
180
Dolomite
Gas
Sand-stone
SALT DOME
TRIASSIC
1,800
Shale
225
2,100
ZECHSTEIN
PERMIAN
Dolomite
Gas
Sandstone
2,700
270
COAL
CARBONIFEROUS

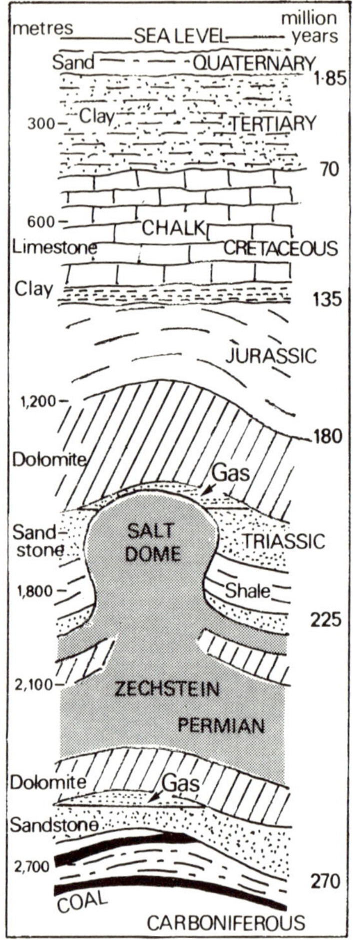

Fig. 92

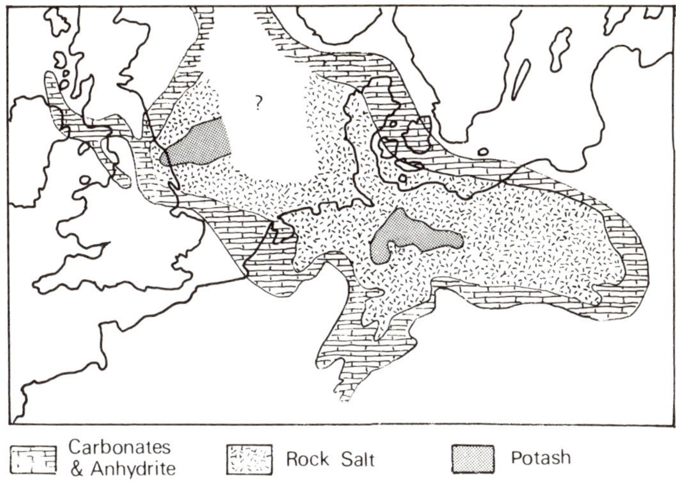

Fig. 93

Figs 92 and 93 The rocks beneath the North Sea and surrounding areas contain several kinds of valuable mineral deposits. Nowadays hydrocarbon fossil fuels and evaporate salts are exploited in several areas. In the North Sea, hydrocarbon fossil fuels are found in association with the Permian Zechstein deposits. Fig. 92 A typical section through the rocks lying beneath the North Sea showing trapped gas deposits. Note: Evaporate deposits are usually an important feature of the rock strata in oil and gas fields. Rock salt is impervious to oil and can form a cap rock which prevents oil and gas in lower strata from escaping to the surface and becoming lost. As salt is less dense than most sedimentary rocks, it buckles in some places and flows upward to form mushroom-like domes which aid the accumulation of oil and gas. Apart from trapping fossil fuels, evaporate salts are valuable for use in the chemical industry, particularly potash (K_2CO_3). Fig. 93 The North Sea and surrounding areas showing the known extent of Permian evaporate Zechstein deposits.

pressure, or fixed at ambient pressure, or mobile at ambient or atmospheric pressure.

Up to the present time all commercial oil drilling involving hole-depths of more than 3,000 to 4,500 metres (10,000 to 15,000 feet) has been done by free platforms or ships. Holes up to a depth of 30 metres (100 feet) can be drilled from small remote-controlled rigs placed on the sea floor, and these have been used for mineral sampling surveys and foundation engineering.*

In shallow water operations, divers using traditional equipment inspect, maintain and service drilling platforms. In deep water submersibles are now taking over completely.

From a single platform ten to twenty wells are usually drilled, slanting outwards to tap the largest possible area of the field in which they are working. Pipelines connecting each platform gather the oil which is then led to a central point for transmission through a major pipeline to the shore.

Exploratory drilling at sea began from traditional vessels modified for the purpose, and much modern exploratory drilling is still accomplished in the same way. When an oil field has been proved, the drilling ship is replaced by a fixed platform resting on the bottom. Drilling from vessels is less expensive than from fixed platforms. Ships are much more mobile. Traditional fixed platforms need to be towed to the operation site, and journeys may take several weeks. In addition insurance rates for platform drilling rigs under tow are extremely high.

One of the earliest innovations of the platform drilling rigs was the development of the floating jack-up rig, also sometimes referred to as an elevating barge. It may have anything from three to a dozen legs which stick through sleeves in a barge hull, and when under tow they can be carried clear of the sea bottom. On location, the legs are pushed down on to the bottom and can be adjusted to raise the platform clear of the waves. The first jack-up rig made its appearance in the Gulf of Mexico in the late 1950s. The rigs are now designed to operate in sea depths up to 90 metres (300 feet), and they provide a wave-top platform clearance of over 20 metres (70 feet). Fixed plat-

* Already fully automated probes can obtain such samples in up to 2,000 metres of water and then return to the surface under their own buoyancy.

form rigs, constructed of large tubular frameworks pinned to the seabed with piles, can operate in depths up to 120 metres (400 feet) or more.

More recent mobile drilling platform rigs are semi-submersible vessels which are in effect modified versions of the more traditional drilling vessels. The semi-submersible platforms provide greater stability for drilling in heavy weather. But the very newest platform rigs are self-propelled, semi-submersible vessels such as the *Ocean Prospector* built in Japan, which measures 103 × 79 metres (344 × 264 feet) and has been designed to drill to 7,500-metre (25,000-feet) depths. It is formed of four submersible pontoons supporting sixteen columns and is able to travel at speeds up to seven knots forward and three knots in reverse (see Fig. 46).

When oil exploration goes much deeper,* it will become much cheaper to install production valves and gather pipe-lines from the well-head on the sea floor. Drilling rigs may also be built on the sea floor, but with the development of surface vessels of high stability it seems likely that drilling will still be accomplished from the surface.

Invention and Innovation

One of the challenges of modern oceanography is the development and invention of new devices and techniques in order to exploit the oceans' resources. As stated already, there are numerous parallels between the problems in outer space and those in hydrospace. In particular, without the modern development in micro-electronic technology it is doubtful if exploration in either realm would have progressed as far as it has today. The electronics industry provides the ways and means of control and observation where man cannot control and observe. The blind underwater groping of yesteryear has been replaced by television and a whole complex of electronic gadgetry.

The development of sonar has provided man with an important underwater probe with many applications. All early experiments with sonar were little publicized because of its military importance. Nowadays its civil and research applications as an underwater probe

* For example, *Glomar Challenger* has discovered oil-bearing shales at great depth.

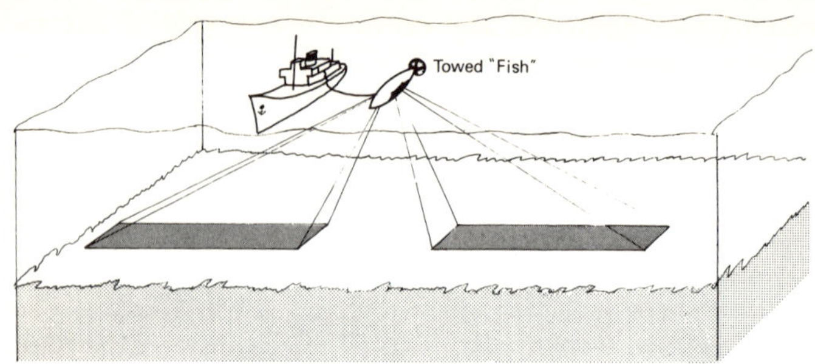

Fig. 94 Side-scanning (side-looking) sonar is now a powerful new underwater tool. Pulsed acoustic signals are radiated by transducers carried in a towed 'fish'. The beams intersect the sea floor where energy is then back-scattered. As the ship progresses on its course, the back-scattered echoes are recorded permanently and continuously on a paper strip recorder revealing the nature of the sea-floor geology in a remarkably vivid way. Using this technique, undersea geological mapping can be carried out in a similar way to aerial photography on land surfaces. Routes for underwater pipelines can be planned with great accuracy; in addition there are many other technological and biological applications.

are numerous. Sonar can be employed in fish hunting, fish counting and navigation, and in marine geophysical and geological investigations. Side-scanning sonar, using beams at right angles to the ship's track, is one of the most important geological bottom-mapping tools at present in use. Side-scanning sonar is also one of the best methods of surveying new seabed pipeline routes. By using high resolution sonar, it is possible to recognize the nature of unseen targets by their characteristic patterns of sonar echoes, thus producing pictures in sound in situations where television cameras cannot operate usefully.

Electronics have also provided the means of submarine wireless. Nowadays, using electrical pulse transmission, underwater communication between vessels can take place over distances of up to 50 kilometres. Automatic weather stations will soon replace more traditional weather ships; they are much less expensive to operate, less hazardous to human life and can remain in commission for long periods without maintenance.

Contemporary marine technology has a wide spectrum of new

invention and innovation. For example, a new kind of electro-magnetic underwater shark barrier has been developed, which has already been erected near Durban in South Africa. Pulses pass through a cable on the sea floor, and these are transmitted to sharks above, causing them intense pain. Detailed research has revealed that sharks are vulnerable to particular electro-magnetic pulses to which humans fortunately remain immune. Another new anti-shark device is the 'Shark-Dart', which is claimed to be the safest and most effective defence against sharks and other marine predators. The 'Shark-Dart' can be used as a dagger or as a lance and loaded with a CO_2 cartridge; the impact releases the CO_2 gas which expands, rupturing the shark's internal organs and forcing it to the surface. Its main advantages over other methods of killing sharks is that it is noiseless and bloodless, killing one shark will not attract others, and the weapon can be reloaded underwater very quickly.

Another innovation employed in sea-floor exploration is nuclear prospecting for minerals – using the uranium isotope, californium-252. The isotope has a half life of $2\frac{1}{2}$ years. Neutrons emitted during its decay are absorbed by minerals which then in turn emit characteristic gamma-rays. Californium-252 is a good isotope source for underwater prospecting, since it is compact and therefore highly portable, and at present it is used in probes which may be operated from either surface vessels or submersibles. Sea-floor mineral deposits can be calculated very accurately simply with a sensitive germanium-lithium detector sealed in the probe head. Results so far reveal concentrations as low as 3 parts in 100,000. Its application to metallic elements and minerals is very wide, including gold, silver, uranium, vanadium, copper, aluminium, fluorine, tin, silica, sodium and iron.

For construction and mining operations undertaken in shallow depths, bulldozers and excavators have been designed for excavating foundations and for concentrating mineral deposits. They may be operated from the surface by remote control, and thus do not involve men working at ambient water pressures.

P.L.B.

10 DESALINATION

Sea-water yields fresh water by natural or artificial distillation and condensation. As we have seen, about 97 per cent of the world's stock of water is in the seas and oceans, leaving only about 3 per cent as fresh water, or as ice at the poles.

As human populations increase, and as their standard of living rises and calls for ever more fresh water, this is likely to be one of the first of the essentials to run short. Already the present supplies of fresh water may be used several times over before being returned to the sea, and already new water storage reservoirs are becoming more difficult to site, and more expensive to build. So the economic desalination of sea-water to supplement natural supplies of fresh water is an urgent problem.

(*a*) *Small-Scale Desalination*

There has been much interest in the supply of fresh water for men cast away in lifeboats or rafts after wreck or war casualty. The amount of fresh water to be got by distillation of sea-water is little more than the volume of fuel used, and the ion-exchange resins, such as are used in domestic and industrial water-softeners, are not effective against full-strength sea-water. If some fish can be caught, these can be chewed by thirsty men, for their flesh is much less salty than sea-water.

The heat of the sun can be used to evaporate sea-water, to be condensed as fresh water. Salt water is run into wide but shallow tanks with the bottom painted a matt black. Over them are placed sloping roofs of clear glass. The sun, shining through the glass, falls on the tank bottom, which absorbs the heat and evaporates the water, which then condenses on the sloping glass roof and runs into a container. Several pints of fresh water a day can be made in this way.

A pilot-scale installation on the island of St Croix in the West Indies is planned, for condensing fresh water from the trade winds. Cold ocean water will be pumped up from several hundred metres depth, where its temperature is about 7° C. (44° F.), and circulated through condensers placed in the path of the warm and moisture-laden trade winds. A significant volume of water could be condensed; and the deep water, now warmed somewhat in the condensers, can be used for fish farming trials (see Fig. 70).

(*b*) *Industrial Desalination*

On the full commercial scale, there are two practical methods of desalination, namely distillation and freezing. Distillation calls for about 555 calories per kilogram (1,000 Btu per pound) of water evaporated, whereas freezing calls for only one-seventh of this energy input, which can, moreover, be got wholly from a standard electricity grid supply.

There are many sea-water distillation plants, and even many years ago the daily world production of desalinated water exceeded 455 million litres (100 million gallons). Some of the plants are small, like the one on the island of Guernsey, which is operated at times of water shortage to supplement the natural supplies. Some are large, like the one which supplies the American city of Key West with 9 million litres (2 million gallons) a day for the 34,000 inhabitants, who are now independent of the mainland for their water; or very large, like the plants which supply Kuwait and Abu Dhabi on the desert coasts of the Arabian Gulf (see Fig.71).

Nor are distillation plants a novelty. When Suakin, on the Red Sea coast of the Sudan, was the base for military operations against the Mahdi in 1884, water had to be supplied to thousands of people in a land without fresh water. The massive boilers and chimneys of the distillation plant still stood there until recently, for the very dry air did little harm to the plant. No doubt it was fired by coal; that was the heyday of Welsh steam coal.

But coal is not an appropriate fuel nowadays, and the world's greatest sea-water distillation plant is at Kuwait, where fuel in the form of waste natural gas costs nothing since it has to be burnt away in any case. It is indeed lucky that oil has been found in so many of the world's desert places; for the big populations attracted by the prosperity and spending power of the oil industry can now be supplied with abundant fresh water.

With a bountiful supply of fuel, Kuwait plans an installed sea-water distillation capacity of 2,727 million litres (600 million gallons) a day by 1985. The sea in the Arabian (formerly Persian) Gulf is both warm and salt; the distillation plant takes in sea-water at a salinity and temperature respectively of 43 parts of salt per thousand and 32° C. (90° F.). Each 7 million litres (1·5 million gallons) of this water yields by distillation 5 million litres (1·1 million gallons) of boiler-standard distilled water containing only a trace of salts. The brine, concentrated by distillation to 63 parts of salt per thousand, and at a temperature of 40° C. (105° F.), is discharged back to the sea.

Very efficient flash-boilers are used for the distillation, in which the new sea-water being pumped from the sea is used to condense the steam boiled at 93° C. (200° F.) under reduced pressure, itself being heated in the process. Passing through this boiler in twenty-six stages, 9 kilograms of distilled water are obtained from each 500 calories (10 pounds per 1,000 Btu) of heat put in. The plant is so well automated that it can be operated by a shift foreman, two charge hands and two operators, who also control from a central control room another 45 million litres (10 million gallons) per day. The cost of this water is about 29 pence (73 cents) per 4,500 litres (1,000 gallons) at 60 per cent utilization, and 21 pence (52 cents) at 90 per cent. But these low costs owe much to the cheapness of the fuel; and even so, it is dearer than water from natural sources. In London, for example, water collected, stored, processed and delivered to the customer's premises costs about 13 pence (32 cents) per 4,500 litres (1,000 gallons). But even here, as the easier and cheaper sources of water have to be supplemented by more difficult and expensive schemes, the cost is bound to rise, and then desalinated water may become more competitive.

(*c*) *Freezing*

When sea-water is frozen, the ice formed is pure water; icebergs floating in the sea are mountains of fresh water. Captain Cook replenished his water-casks from floating sea ice on his trips to far southern latitudes. Sea-water can be frozen artificially, and the ice crystals separated, washed free from adhering brine, and melted to give pure water.

The freezing agent is the gas butane, which is in ready supply nowadays. It is easily liquefied under pressure at atmospheric temperatures, and the compression can be done with electric power from the electricity grid. The liquid butane boils at −10° C. (14° F.); so when it is pumped into a 'crystallizer' containing sea-water, the latter warms the butane, turning it back to gas, and in turn is cooled to freezing point by the latent heat of evaporation of the butane.

The mixture of brine and ice crystals passes to a washing tower, where surplus brine is washed off the ice crystals; these are then melted in a vessel in such a way that the latent heat of fusion of the ice is used to cool the recirculating butane gas as it is again liquefied under atmospheric pressure. The process is continuous, and economical in the use of energy.

A notable example of this type of desalination plant is at Eilat, on the Red Sea coast of Israel. A similar plant, with a designed capacity of 4·5 million litres (1 million gallons) of fresh water a day, was intended for the English town of Ipswich, where it was calculated that, over a period of twenty-one years, this plant, operated at a high load factor, would be less expensive than a new reservoir, taking into account all probable ranges of growth in demand, discount rates and power costs. But at present the Ipswich scheme is in abeyance.

(*d*) *Multi-Purpose Schemes*

In the Environment Research Laboratory at the University of Arizona, a combined power/water/food-production process has been devised for arid climates, and has been tested with success in Mexico. A plant on the same lines has been installed on the island of Sadiyat, in the Arabian Gulf off the oil state of Abu Dhabi. This is a diesel-driven power station, the waste heat of which is used to distil fresh water from sea-water at a rate of 318,000 litres (70,000 gallons) a day. This fresh water is partly used for drinking, but also partly to irrigate greenhouses made of plastic. As in the established practice of hydroponics, the water is made to dissolve nutrient salts before being led along perforated plastic pipes among the root-systems of vegetables planted in sand. An underlay of plastic sheeting prevents loss of water by seepage. The greenhouses are kept cool by the evaporation of sea-water, and the high humidity maintained in the greenhouses diminishes the loss of water by transpiration.

Meanwhile, the exhaust gases from the diesel generators are passed through a scrubbing tower, where they are cleaned at a rate of

5·7 cubic metres (200 cubic feet) per minute, and then passed into the greenhouses, where they supplement the atmospheric carbon dioxide in supporting a high rate of photosynthesis. Heavy crops of cucumbers and tomatoes have been grown though, as might be expected, the first crops have been more expensive to grow than those grown locally for the market.

C.F.H.

11 POWER FROM THE SEA

The oceans, with their prodigious volumes of water always in movement, represent a source of energy ready for use, if an economical way can be found. One such way is to use the head of water above mean sea-level at high tide in tidal power stations. The rising tide fills a reservoir, from which the water can be passed through a water-wheel or turbine as the tide falls.

Tide mills are not at at all rare. A survey of the English counties of Devon and Cornwall alone records thirty-two tide mill sites, though none is now in use. The best-known tide mill is at Woodbridge in Suffolk, England, which was built about 1170, and remained in use until recently. Dutch settlers took the idea with them to New York, where a tide mill at Spring Creek was in operation until a century ago.

A tide mill has a large embanked pond with two or more sluices equipped with clapper-gates which would swing open to let water in on the rising tide, but close when the tide began to fall. The impounded water escapes through an undershot water-wheel or, in a more modern form, a horizontal Francis turbine working a generator. But the head of water must vary with the state of the tide, and as the time of high water varies, the mills could not always have been run at the most convenient times for customers. Indeed, it is recorded that many tide mills closed because they could not find men willing to work such irregular hours (Fig. 68).

When a Severn Barrage scheme (at the mouth of the English River Severn) was under discussion, design studies proposed to avoid this disadvantage by using some of the power developed at falling tide to pump water up to a storage reservoir, whence it could fall to work the turbines on a part of the rising tide. One variant of the scheme envisaged 200 prefabricated concrete caisson units, each carrying a 9 metre (30 foot) diameter axial-flow turbo-generator rated at 33 megawatts. There are plans for very big tidal power stations to supply the developing Soviet far north. Planned outputs are 12,650 megawatts in multi-purpose installations. The scheme for the harnessing of the great tidal heads of Passamaquody Bay is at present in suspense, the rapid success of nuclear power making such schemes look less attractive; and indeed, in general, tidal power development has disappointed its advocates.

The Rance Tidal Power Station

One scheme, however, is complete and in full production. This is the work of the French State Electricity Board, Electricité de France, and should be associated with the name of the engineer M. Maubouissin. There is a large tidal range on the north coast of Brittany and the estuary of the River Rance has a tidal rise of 11·4 metres (37 feet) on average, with 13·5 metres (44 feet) at spring tides. The estuary is large, with a flow of water reaching 18,000 cubic metres (211,500 cubic feet) per second at flood and ebb tide; but the inflowing river is small, so that no large load of silt had to be provided against. A small rocky island was conveniently placed to help the construction of the barrage.

The completed barrage is 15 metres (49 feet) high above mean sea level, and encloses a basin with a capacity of 184 million cubic metres (2,166 million cubic feet) between the tide levels of 0 and 13·5 metres (0 and 44 feet). The volume of water available at each tide is four times that of the River Rhône at the city of Avignon; however, the total power obtained is only half that of one of the Rhône dams. But it is an assured flow, unaffected by floods or droughts.

The Rance tidal barrage is unusual in that the thrust of the water is not from one direction only, as with a river barrage, but alternately in opposite directions. The sluice gates had to be of special construction for the same reason.

The barrage includes three sections, namely, a ship lock at the western end, a long central power-house, and a series of six sluice gates to control the flow of water. These gates are each 15 metres (49 feet) wide and 10 metres (33 feet) high, and can each pass 1,500 cubic metres (17,640 cubic feet) per second for each 1-metre difference of level.

The power-house is built over twenty-four concrete channels in the bed of the original estuary, and each channel houses a power bulb consisting of a generator turned by a 9 metre (30 foot) diameter Kaplan turbine, giving 10 megawatts at 94 revolutions per minute. Each of these power bulbs can act both as generator and as pump, in both directions of tidal flow and both directions of rotation, since the turbine blades can be reversed or feathered. Each power bulb is serviced by means of a ladder leading down a shaft from the power-house.

Cycles of Operation

There are so many variations in the ways in which this station can be operated that a computer is used to programme the best results. One such cycle could be as follows:

1 The tide rises, and enters the basin through the twenty-four turbines, generating electricity for the national grid.
2 Near the top of the tide, the turbines are feathered and the sluice gates opened – so as to complete the filling of the reservoir.
3 The sluice gates are shut, and the power bulbs are run for a short time as pumps, drawing power from the grid to pump additional water into the reservoir to give an extra head of 50 centimetres (19 inches).
4 On the falling tide, the power bulbs again generate electricity, but now the 50-centimetre (19-inch) overfilling will give back to the grid twelve times the energy it borrowed.

More power is got from the outflowing water than from the inflowing water at rising tides, and in both cases the power depends on the head of water. At 11 metres (36 feet) head, both cycles give 10 megawatts per power bulb; at 7 metres (23 feet), estuary-to-sea gives 10 megawatts, but sea-to-estuary 9·5 megawatts; at low head of 3 metres (10 feet), estuary-to-sea gives 3·2 megawatts, sea-to-estuary only 2 megawatts.

The annual output is about 500 million kilowatt-hours, and France is rightly proud of this major pioneering engineering work.

Claude Thermal Power Plants

The so-called Claude process uses the difference in temperature between the warm surface water in tropical seas and the cold deeper water (Fig. 69). As long ago as 1931, a Claude power station with a net output of 22 kilowatts was built on Cuba. Just after World War II, a similar plant was built at Abidjan, on the Ivory Coast of West Africa.

The principle is to boil the warm surface water, at about 35° C. (95° F.), under reduced pressure, and condense the low-pressure steam in an annular condenser cooled by deep water pumped up from several hundred metres depth and at a temperature of about 6° C. (43° F.). Between evaporator and condenser is a turbine coupled to a generator.

But this type of power station seems to have had little success so far, and the plant at Abidjan was dismantled in 1961, after less than 20 years' service.

Use of Cold Deep Ocean Water in Condensers

A power station planned for Hawaii will draw its cooling water not from the sea surface water, but from a depth of 152 metres (500 feet), where the water has a temperature of only 7° C. (44° F.). The increased efficiency of the condensers of the power station would effect a net fuel saving of about £148,000 ($370,000) a year, which would pay for the extra cost of the long pipeline. Because of the steep underwater slope of the Hawaiian islands, this need be only about 3 kilometres (1·9 miles) long. In cooling the condensers, this cold deep water, which is also rich in nutrients, will be warmed to about sea-surface temperature and discharged back to the ocean. There it will produce a high rate of plankton growth and might affect favourably some local fisheries. The possibility of using this nutrient-rich deep water to support a fish-farming industry in suitable atoll lagoons is discussed on p. 225. It has been estimated that a large atoll, such as that of Kwajalein in the Marshall Islands, could in this way support the culture of animal protein for 10 million people.

C.F.H.

12 BASIC FOOD RESOURCES OF THE OCEANS

The chief resources of the sea are obviously water and dissolved salts. and these resources are virtually inexhaustible. But some of the biological, self-renewing resources such as fish and whales are not inexhaustible and indeed are suffering from over-exploitation (pp. 231–3). Pollution may threaten the survivors (pp. 235 ff).

As to the water; on the 29 per cent of the globe which is land, plant and animal life depend largely on fresh water, that is water which contains only a very small amount of dissolved salts. By natural evaporation from the sea surface, and condensation on the land, this supply of fresh water is maintained. On the industrial scale, fresh water can be artificially distilled from sea-water, or separated from it by freezing (see 'Desalination', pp. 195–7).

As this fresh water returns to the sea down the rivers, it dissolves a great deal of land material, and this has accumulated over the millions of years during which the oceans have existed (see pp. 41–6).

Dissolved Salts

The sea probably contains all the elements naturally occurring on earth (on the mineral composition of sea-water see also pp. 8–9). Of the 92 elements, 62 have been estimated. Some are present in prodigious quantities, such as sodium and chlorine which are extracted by evaporation for the manufacture of common sea-salt (p. 183), and magnesium and bromine are in sufficient concentration to be extracted commercially (see p. 183). The biologically essential, such as nitrogen, phosphorus, potassium, calcium and iron, are much less plentiful, but in spite of this, plants are able to concentrate them from extreme dilution. Besides these essential elements, there are many trace elements important for life, including manganese, magnesium, molybdenum, copper, boron, zinc, cobalt and many others. These also can be concentrated by marine organisms to many hundreds or even thousands of times the concentration naturally occurring in the sea. If we could find out how they do this, we might be able in the same way to extract valuable metals from the sea. But, as yet, no sea creature has been found to concentrate gold for our benefit.

Organic Matter

As the fresh water flows back to the sea, it also carries with it much organic matter, in the form of carbon compounds of varying degrees of complexity derived from plant and animal life, either in solid form as detritus, or as colloidal and dissolved carbon compounds. In the sea, these quantities are added to on a prodigious scale by the living activities and dead remains of marine plants and animals. A recent estimate suggests that 5 to 10 million million tons of organic matter of this kind exist in the oceans, and this source of potential energy must play an important part in the life of the sea. A large group of humble plants and animals are able to live off this supply of organic matter. Bacteria are not common in the sea as free-living organisms, but are very plentiful on any substrate such as particles of detritus. The detritus fragments, enriched by their population of bacteria, become nutritious food for filter-feeding organisms. In fact, the detritus, and the dissolved and colloidal organic matter, form a separate and important food chain (see p. 212), distinct from that based on photosynthesis, about to be described.

This production of 'bacterioplankton', at the expense of dissolved organic matter, may in fact greatly exceed production of new organic matter by phytoplankton, in the ocean. Feeding experiments have shown that filter-feeding animals such as copepods, cladocera and mollusc larvae among zooplankton, and sponges, worms, oysters and coral polyps among bottom-dwellers, are able to feed on planktonic bacteria.

In very deep water, far below the influence of sunlight, minute plants have been found growing in a healthy manner, such as the coccoliths of the deep Mediterranean, and even diatoms and blue-green algae. On the sea bottom at very great depths, a surprisingly large fauna of worms and shellfish lives in the mud, which also has a considerable population of bacteria, showing that enough organic matter is present to support a substantial fauna and flora. Not all this organic material is small, for large pieces of waterlogged wood, fruits, etc., have been found on the ocean bottom in great depths. These may be well colonized by boring animals which are a source of food for other predatory animals. The volume of living animals per square metre in these great depths is correlated with the abundance of organic production in the lighted surface layers of the sea above, again showing that the bottom fauna, even in great depths, is ultimately fed by the rain of organic matter falling from the sea overhead, the richest areas for surface production being also those of comparatively rich abyssal fauna.

It is known that diatoms can grow on dissolved organic matter, and so can the brine-shrimp *Artemia* and probably many copepods. A whole group of deep-sea primitive worms, the Pogonophora, have no gut at all and they live by absorbing dissolved organic matter over the whole body.

So this massive reserve of organic matter present in the sea is significant; and since it includes such substances as vitamins and growth-promoting substances, its presence in sea-water makes for an even more perfect nutrient medium than its content of inorganic nutrients such as phosphates and nitrates would suggest.

Photosynthesis

All this nutrient material, organic and inorganic, is distributed about the oceans by the current systems, awaiting its call to return to the cycle of life. This comes through the second of the great resources of the sea, namely sunlight falling on the surface. This source of energy provides both heat and light. The heat warms up the surface, providing at first a favourable temperature for life processes; but also able to form a layer of warm, light, water, which may hamper the full use of the nutrients available.

In the sea, as also on the land, the light energy of sunlight is intercepted by the green pigments of plants.

Though the larger seaweeds we know do play a part on the sea shores, by far the most important part in marine photosynthesis is done by microscopic plants, the diatoms and peredinians chiefly, and also by ultra-microscopic flagellates called *nanoplankton*. Though even now not well known because of great technical difficulties in examining and classifying them, it is known that they may be several times as important as all the so-called 'net plankton' combined.

Ultra-minute as they are, they are also an important source of food for many filter-feeding animals

The energy of sunlight intercepted by the green pigments of these plants is used to bring about the reduction of dissolved carbon dioxide to carbon, and thence to carbohydrate. There is never a shortage of carbon dioxide in the ocean, for the abundant bicarbonate ions can split off carbon dioxide as soon as free dissolved carbon dioxide begins to be used up.

The reaction, probably the most important in the world, can be stated as follows:

$$6CO_2 + 6H_2O = 6C_6H_{12}0_6 + 60_2$$

This reduction is made possible by the Sun's energy; it will not take place in the dark, nor in the absence of chlorophyll. Because energy has been put into the reaction, the product, carbohydrate, contains potential energy which can be recovered. When sugar burns, which it does fiercely, it is giving back the energy stored at its formation. But in the plant, the energy is recovered slowly, under control by the action of enzymes, to build up larger and larger molecules, until at last new living matter is formed. The whole complex process has been summarized as follows:

1·3 million calories of solar energy, + $106CO_2$ + $90H_2O$ + $16'NO_3$ + $1''''PO_4$ + traces of mineral elements	=	13,000 calories of potential energy represented by 3258 grams of protoplasm (106C, 180H, 460, 16N, 1P, 815 gm mineral ash) + 154 0_2 + 1,270,000 calories dispersed as heat.

Evidently, the process of photosynthesis is not efficient, for only 1 per cent of the available solar energy is fixed in new living matter, while the remaining 99 per cent goes to warm the sea surface, where it may create new problems.

Amount of Plant Production in the Sea

(a) Total

Production of new plant material, on which all other life depends, varies widely from region to region, and at different depths.

As the light passes down through the water, it becomes weaker, and also changes in colour, as it becomes scattered and absorbed by the water and the particles suspended in it. At a certain depth, which depends on many factors, the light can only just produce enough oxygen to balance the loss of oxygen due to the respiration of the plants and animals in the water. Below this depth, there is a net loss of oxygen.

But the principal cause of the variation in the rate of plant production, from one area of sea to another, is the amount of nutrient salts present, and these vary in both time and space. Comparative maps show a good agreement between the oceanic areas of abundant nutrient salts and those of rich phytoplankton (Fig. 56).

Various methods have been used to attempt to determine the gross rate of primary plant production in different localities and in different areas of sea.

Atkins, at the British Plymouth Laboratory, obtained an estimate of 250,000 kilograms of dextrose sugar per square kilometre for the first six months of the year in the western English Channel, by observing changes in alkalinity (as carbon dioxide is removed from the water in the course of photosynthesis, it becomes less acid). A second estimate in the same region, based on the rate of assimilation of phosphate, gave an annual wet-weight crop of diatoms of 5·4 tonnes per hectare (13 tons per acre) – roughly the same as the production of a potato field. In another series of experiments, using the known relationship between carbohydrate fixing and oxygen production to obtain estimates of plant growth at different depths and locations by measuring the net changes in the level of dissolved oxygen in the water (taking account of the consumption of oxygen by respirating organisms also present in the water), annual rates of production of from 5 to 1,000 grams per square metre were obtained, varying with depth and from region to region, the highest rates being found in coastal waters and the lowest in the gyres in the open ocean.

The most accurate results however are obtained from a method pioneered by the Danish biologist Steeman Nielsen on the 1950–2 world cruise of the *Galathea*. In this method a quantity of sea-water from a given depth is enriched with a radioactive isotope of carbon which plants use in the same way as the normal carbon they obtain from atmospheric carbon dioxide; and the amount of carbon fixed as carbohydrate in a given time and under the conditions of temperature and sunlight prevailing at that depth (either simulated or achieved by lowering the water in a bottle) is calculated by measuring with a Geiger counter the β-emanations from all the plants in the water able to carry out photosynthesis, collected by filtering the water through an extremely fine membrane filter.

Nielsen's results reflected a somewhat smaller variance between coastal and open-ocean rates – 0·55 to 0·62 grams per square metre daily near the coasts of California and New Zealand, for instance, and 0·13 to 0·16 grams per square metre in the open Pacific.

The annual plant production of the seas and oceans of the whole world was estimated by Nielsen at 12–25,000 million metric tons of carbon, equivalent to about 40,000 million tons of organic matter and roughly equal to what the land produces in two-fifths of the sea's area. More recent work accepts a figure of 15–20,000 million metric tons of carbon fixed annually in the seas and oceans.

(*b*) *Seasonal*

In temperate seas, where there are marked seasonal changes in temperature, the surface layers of the sea become warmed up, as shown in the formula on page 205, and as the summer progresses, this layer of warmed surface water, which is lighter than the cooler, deeper water, becomes thicker, so that the water-column develops a stability, with the warmer, lighter water at the top and heavier, cool water below. This stability can only be overturned by a great deal of energy. In spring, while the warm surface layer is still thin, strong winds can overturn it; but it would take a full gale in the summer, and even so, the stability would soon be restored.

But, as the summer draws on and the light grows stronger, the plant plankton soon uses up the nutrients present in the warm surface layer, so that production may come to a halt, even though other factors such as light and dissolved carbon dioxide are in ample supply (Fig. 54). Thus we may have the paradox of poor production in the surface water, while below in the deep cool water there are plentiful nutrients. Yet the stability of the water column prevents this richer bottom water from rising into the lighted zone.

In the autumn, the sunlight begins to wane, and the temperature of the sea surface begins to fall. By mid-autumn, the surface water will have cooled to the point where the stability of the water column is small enough to be overturned by strong winds and gales. The resulting admixture of rich bottom water brings about a new outburst of plant plankton growth, though less intense than that of spring, for now the days are shortening. Finally, in the winter, plant growth virtually stops, and the winter gales bring about a complete mixing of the water-column. Therefore, when the sunlight strengthens once more in the spring, there are plenty of nutrient salts in the water to begin the cycle over again.

This is a simplified picture of the annual cycle of temperate seas. Fish which feed on the plant and animal plankton, such as the sardines, pilchards and menhaden, reflect these seasonal changes. In Fig. 58 the average monthly mean weight of food in the stomachs of large samples of pilchards is shown; there are two separate periods of more intense feeding, namely, in spring and autumn, and these include a high proportion of diatoms and other minute plants, so much so that the colour of the stomach contents may then be a bright green. But in the summer, the plant plankton becomes scarcer and the animal plankton more plentiful because, as said above, the plants may use up all the nutrients in the surface layers of the sea, while the copepods, which are grazers on the plants, show a very rapid increase in quantity and graze down the plant plankton. In the summer, the stomach contents of the pilchards become reddish in colour.

In the high latitudes of the Arctic and Antarctic Oceans, daylight is continuous during the summer months, but the sun's heat is not enough to form a strong stratification in the open surface waters. So there is a very rapid and continuous use of the nutrient salts in the open waters, and a very dense growth of plankton. This supports, through the food chains, a slow-growing but rich marine life including, in the Northern Hemisphere, abundant fish which support some of the world's richest fisheries off Iceland and Greenland and in the Barents Sea: and, in the Southern Hemisphere, the great whale fisheries. In fact, in the open waters of the Antarctic, there is so much nutrient material that even great outbursts of plankton growth fail to exhaust the supply; the Antarctic surface water, moving north to sink below the sea surface at the Antarctic Convergence (see Fig. 26), carries a rich cargo of nutrients which, when brought to the surface again by water movements, give rise to the world's greatest surface fisheries, off Peru and South Africa.

(*c*) *Regional*

In the tropics, where there are no warm and cold seasons and the ambient temperature is always high, a superficial layer of warmed water, a couple of hundred metres thick, is a permanent feature, and here the sea may have the deep indigo blue colour of oceanic badlands. Exhausted of its nutrient salts, and sometimes an area of downwelling, the sea can there nourish only a small plankton population. The general level of organic production in the open oceans seems to be about 0·1 grams of carbon per square metre per day (gmC/m^2/day), but in the Sargasso Sea it is less, about 0·04 gmC/m^2/day.

Beneath this thin superficial warm layer, which absorbs almost all the incident sunlight, lies the immense depth and volume of the deep ocean, very cold, pitch dark and rich in the nutrient salts, chiefly nitrates and phosphates which, as we have seen, marine plant-life, like the more familiar terrestrial plant life, needs for full development in the presence of sunlight and dissolved carbon dioxide.

In situations where this nutrient-rich deeper water is brought to the surface, tropical waters can show much higher productivity. Where regular winds blow offshore, the spent surface water is blown to sea, and nutrient-rich deep water is dragged up to replace it. The work done in lifting the cold and heavy deeper water is provided by the energy of the wind. Two famous regions of such upwelling are off south-west Africa, and off Peru (see Figs 55–7). The *Galathea* found production values of 0·5 to 0·8 gmC/m^2/day off south-west Africa, with one value of 3·8. Off Peru, similar values have been observed, namely 0·4 to 0·8.

Richer areas are also found near and over submarine banks, and along the margins of the continental shelves; for in such places turbulence and eddies bring up the rich deep water. This is responsible for such rich fisheries as those of the continental shelves. Finally, where two ocean currents flow past each other, strong eddies are set up, and the great Newfoundland Banks fisheries are one of the most striking results of the upwelling of the rich deep water there. In the open oceans, where the North and South Equatorial Currents run beside the Equatorial Counter-Current, eddies are set up which result in a belt of productive water along the equator. The *Galathea* found values of 0·26 and 0·5 gmC/m^2/day in this belt.

Where these areas of richer production occur, there are the rich fisheries. If a greater part of the oceans could be made to produce, say, 0·2 instead of 0·1 gmC/m^2/day, a world protein shortage would be postponed for some years. Clearly, enormous amounts of energy would have to be applied to secure an artificial overturn. But with the advent of atomic energy on the largest scale, the possibility is less remote than it seemed at one time. On the small scale, this is being tried (see p. 225).

Though by far the greater part of marine production is done by the phytoplankton, including especially the nanoplankton, other marine plants may have local importance. The fixed red, green and brown seaweeds break down to detritus, on which a whole fauna subsists. *Zostera* and *Poseidonia*, the turtle- or eel-grasses, may grow luxuriantly in coastal waters and estuaries, and be an important local source of food material. Important fisheries in Danish coastal waters and fjords are based on *Zostera* and its breakdown products.

Food Chains

The plant plankton is food for a great variety of small crustaceans, especially copepods and Euphausiid shrimps, and for many important molluscs, such as oysters and mussels. A great many invertebrates such as worms and shellfish have young stages which live on the plant plankton for the first few days of their lives. Plankton-feeding fishes are also to a great degree feeders on animal plankton, as Fig. 60 shows. In the sea, as on the land, larger animals feed on smaller, and thus there is a food chain through which the newly formed plant material is eaten by small creatures, and these by larger ones, until we come to the largest fish and whales.

But at each stage of the chain there is a great loss of material, a loss usually estimated at 90 per cent. Animals feed for two purposes, maintenance and growth. Growth, involving an increase of weight, can occur only after the needs of the animal for energy and repair of its tissues are satisfied. It seems that about 90 per cent of the energy gained from the food goes for maintenance, leaving only about 10 per cent for growth. This loss happens at each stage, as may be seen below, where the total primary production in the sea is taken as an annual 16,000 million metric tons of carbon. This total is reduced down the food chain as follows:

Phytoplankton	16,000 million metric tons
Zooplankton	1,600 million metric tons
Eaters of Zooplankton	160 million metric tons
Predators	16 million metric tons

In agreement with this table, the fish which feed on zooplankton are the most abundant; the herring family, for example (anchovies, herrings, sprats sardinellas, pilchards, shads), provide over 40 per cent of the world's supply of fish. It is also significant that some of the world's largest animals, with the greatest food requirements, such as the baleen whales, the whale shark and the basking shark, and the giant manta ray, or devil-fish, which may measure 8 metres (26 feet) across and weigh 2,032 kilograms (2 tons), are zooplankton feeders.

The predatory fish, such as the members of the cod family and its relatives, provide only about 12 per cent of our fish, again as to be expected from the table.

The total weight of sea fish caught in 1970 has been estimated at about 60 million metric tons; if about 10 per cent of this is carbon, then out of the total marine annual production of 16,000 million tons of carbon, only some 0·038 per cent comes to us as fish, though locally higher proportions may be achieved. As to the predators, those which occupy the fourth stage of the table above, there is good reason to believe that the present world catch is nearing the total which can be taken on a practical commercial basis (see p. 227). It would therefore obviously be far more economical to take our seafood at the second stage in the table, where it would be 100 times more plentiful; but so far it has always been cheaper to let whales and fish, which have special aptitudes for this, catch the plant-eating organisms for us (see p. 228).

Such is the conservatism of human tastes, that only a few dozen out of the 25,000 known species of fish are extensively used as food.

The Nourishment of the Animals in the Deeper Water.

Though new organic matter can only be synthesized in the sea in daylight, and in the uppermost 100 metres or so, where the light is strong enough to fuel the reduction of dissolved carbon dioxode (p. 205), life exists in the oceans from the surface down to the deepest oceanic trenches. While the inhabitants of the upper photic layer of the sea, as has been said on p. 210, can feed directly on the phytoplankton and zooplankton, those living deeper down depend for their food on two main indirect sources. These are, firstly, the fall of particles from the photic layer and, secondly, the vertical migrations of animals between the productive surface layers and the deep sea.

On the continental shelves, abundant material reaches the sea bottom by both routes, since the depths are not more than about 200 metres (660 feet), and nourishes the world's most important bottom-dwelling food-fish. The material which falls from the surface includes the dead bodies of plants and animals and the faecal pellets of the grazing zooplankton. It is taken up by filter- and detritus feeders, such as anemones, sea-squirts, tube-worms, lamellibranch molluscs, prawns and shrimps; and these are food for fish.

Over the deep oceans, this material, as it slowly falls, may be eaten by filter-feeding animals of all kinds, and be passed on, as faecal pellets much diminished in food value, to be perhaps enriched by bacteria and eaten yet again by filter-feeders still deeper down in the ocean. These filter feeders, most importantly crustacea, are food for midwater fishes and squids (Fig. 63). Naturally, the farther away from the productive surface layer, the less the material available to support animal life; until, finally, the little indigestible material that is left reaches the ocean bottom, and there joins, especially near the continents, organic detritus derived from the coasts. Thus the deep sea muds may contain enough organic matter, and bacteria, to nourish a fauna of worms, crustacea, molluscs, brittle-stars and sea-slugs (Fig. 62), which supports a sparse fish-fauna (see pp. 203–41).

Probably more important is the transfer of material brought about by vertical migrations. Deep-sea fish, and especially deep-sea eels, produce millions of larvae which rise to the surface of the ocean and grow there at the expense of the abundant food material. When they metamorphose into young eels, they begin the descent to the abyssal depths which will be their permanent home. As they descend, they are very acceptable food for deep-sea predators, and in this way they carry important amounts of food material down into the deep sea. Also, many species of planktonic animals live deep down by day, but migrate upwards into the surface layers at night. There they feed well on the material which has been produced by the phytoplankton during the hours of daylight. When, at dawn, they migrate downwards again, sometimes as deep as 700 to 1,000 metres (2,300 to 3,300 feet), they carry down with them, in their bodies, very large amounts of potential food for deep sea predatory fish and squids, which may ascend from very deep water. This diurnal migration, in which the most important animals are the small bristlemouth and lantern fishes, and the euphausiid crustaceans, is so marked that the layer of migrating animals may often reflect the beams of an echo-sounder, giving the so-called deep scattering layer which can be seen to rise to the surface at night, and descend deep by day. (See also pp. 69–71.)

The sperm whales feed, when over the continental shelves, by diving for bottom-living fish but, when living in the open ocean, dive as deep as 1,000 metres (3,300 feet) to hunt for the big squids (Fig. 61) which are probably feeding where the descended vertical migrants and their deep-sea predators are. As these whales are big animals, with large food requirements, there must be plentiful food at that level in the ocean.

There seems to be some mystery as to how the sperm whales hunt and catch their squids. At several hundred metres below the sea surface, the daylight is very faint; so the whales must use some form of sonar, as the smaller dolphins have been proved to do (see p. 228). Moreover, the squids are the swiftest of all the invertebrates, with their perfect form of jet-propulsion, by which they can dart along at 15 knots (790 centimetres per second). But the sperm whale can only do about 8 to 10 knots. Yet, in the oceans, their food consists largely of squids, and they frequently carry the scars of their fights with the squids they have seized; so they must have some kind of advantage over their swifter prey.

C.F.H.

13 FISHING AND FISH FARMING

The fisherman's livelihood depends on knowing where remunerative concentrations of fish are likely to be found. To a large degree, he draws on his own and other men's experience of past events, where good catches have been taken at certain places and times. But there are also direct means of locating fish, for example, by observing the fishing and diving of seabirds, phosphorescence at night, and discoloration of the water. The echo-sounder, originally designed to give an instantaneous reading of the depth of water beneath the ship's keel, also gives echoes from the bodies of fish; as the sensitivity of the echo-sounder has been improved, its importance as a fish finder has, on fishing vessels, equalled its usefulness as a navigational instrument. Moreover, the actual species of fish can sometimes be identified; for example, herring shoals give a notably different echo-trace from pilchards, and of course the echo-trace will also give the distribution of the shoal in depth.

More recently, the 'asdic', developed during the war for the detection and location of submarines (now known as 'sonar'), has been adapted for finding shoals of fish.

Shoals of fish can be searched for in the horizontal plane, and some sonars have a range of several hundred metres. Whales can also be hunted with the aid of sonar, and a new fishery for herring and mackerel has been developed by the Norwegians in the northern North Sea, locating the fish by sonar, and catching them in purse seines.

Before the advent of the echo-sounder and sonar, a fishing boat on the look-out for herrings would often tow behind her a piano-wire attached to a heavy weight. The vibrations of the wire as it passed through a shoal of fish could also tell an experienced skipper how large the shoal was, and its distribution in depth.

In some more primitive fisheries, master fishermen can listen for fish, since many kinds of fish and prawns make characteristic noises which can be recognized by a trained ear. By thus knowing their distribution, the master fisherman can direct the fishing craft on to the fish by suitable signals.

Attracting Fish for Capture

(a) *Bright Lights*

Many species of fish, squids and prawns are attracted to a bright light shone into the water at night. Not only are they attracted, but they seem to become bemused and incautious. More expensive fishing outfits have clusters of bright lights on booms, and with searchlights also on the ship will attract big shoals alongside, which can then be surrounded by a purse seine. To prevent a loss of light by reflection, the lights may be lowered into the water in insulated water-tight jars. On the smaller scale, the outfit may consist of two small boats each carrying two or three incandescent oil lamps on a bracket on the stern, and a larger net boat. After a longer or shorter period, small fish such as anchovies or sardines will gather in the brightly lit circle, and not fish alone, for the light may attract fast-swimming segments of bristle-worms with crimson bodies and golden bristles, squids which dart about ejecting a puff of ink at each jump and many small fish which may even jump at the lights. Once a good shoal of fish has been attracted, the two light boats row together to mix their shoals, and then one covers its lights and rows away. The remaining boat is surrounded by a purse seine shot by the net boat. When this has been done, this light boat also covers its lights, leaving the fish surrounded by the net. Both light boats pass over the headrope of the net, and uncover their lights to attract another shoal.

(*b*) *Fish Shelters*

Very common all over the world is the use of shady shelters to attract fish. The shelter may be a bundle of big palm leaves, buoyed and weighted; there may be dozens of these shelters in a small area, and the fishermen go from one to the next until they find one which has many fish gathered under and near it. In a modern form, such shelters may be made of condemned motor-car bodies or masses of worn-out tyres. These artificial reefs are often used to improve a sport fishery, and of course the preference of fish for wrecked ships is well known.

(c) *Ground-Baiting*

Ground-baiting or chumming is also a world-wide device for concentrating fish for capture, on both the sport and the professional scale. For example, a beach may be baited at night with fish scraps. As the tide rises, prawns, which are nocturnal feeders, gather in the baited areas and can be caught with cast-nets. Chumming for tunas is on a far larger scale. The tuna clippers have a large bait tank on the stern which has a sea-water circulation, and in this tank are kept alive small fish such as sardines or silversides. When a tuna school is sighted, live bait is thrown out rapidly to attract the tuna to the side of the clipper. The crew then begin to fish with feather jigs dropped among the milling fish. It becomes possible to stop ground baiting, and to keep the tunas in a state of frenzy by squirting strong jets of water among them. The sight of a tuna clipper in full fishing is one of the thrills of the fishing industry (see Fig. 83).

Baiting to attract fish is universal. Crabs and lobsters are caught in baited traps (Fig. 90), and so are the valuable crawfish, though crawfish can be caught in unbaited lengths of pipe or other good shelter. Traps to catch fish may or may not be baited. Like the shelters mentioned above, traps are a place in which to hide, and seem to be as effective unbaited.

Baited Hooks

Baited hooks are adapted to fishing on the largest scale. The use of handlines, as in the Newfoundland and Lofoten cod fisheries, is less important now, though 'banking' still provides some high-quality fish. More common are the 'longlines' which are used for the catching of bottom-dwelling fish such as halibut, skates and rays in Europe, and halibut in the North Pacific. These are heavy lines with hooks at intervals of about 6 metres (20 feet), and as many as 5,000 hooks might be shot by one longliner. The bait used is usually cheap fish or squid, and the lines are hauled by a mechanical linehauler.

The Japanese have modified this to fish at the surface, or in the upper hundred or so metres, for the oceanic tunas (Fig. 84). As with the deep-sea longline, the lines are coiled in baskets, which are joined end to end as the lines are shot. There is a hook every 10 metres (33 feet) or thereabout. The bait used is the saury mackerel, and no other bait seems quite so effective. The line is suspended below the sea surface from buoys, at whatever depth the tunas are likely to be swimming, and this again may be found by hydrographic observations. The boundary region between the warmer surface water and the cooler deeper water is a likely place to try. The line is hauled by motor winches, and though the rate of capture may be low, say from three to six fish per 100 hooks, as many as 7,000 hooks may be shot, and the tunas are big fish; so the total weight of fish caught may be impressive. These Japanese longlines are worked the world round, chiefly along the region of the equatorial currents and counter-currents, where the fertility of the sea is locally greater (see p. 210). Refrigerated mother-ships call to collect the fish and bring new supplies of frozen bait, and the liners spend long periods away from home.

There are very important high-seas line fisheries for salmon.

Fixed Fishing Gears

Many important fishing gears are fixed, so that migrating fish swim into them. One such is the type of net called a 'pin seine' in Guyana, which is also used under various names in other countries. At low tide, a long line of tall stakes is driven into the mud, facing shores which are visited at high tide by valuable fish. A long net is attached to these poles but trampled into the mud at low tide. As the tide rises the fish pass over the net, but at or near high water the headline of the net is raised by the pin seine boat and attached to the tops of the poles. When the tide falls, all fish inside the pin seine are trapped and can easily be picked up.

Then there are the fish-fences of brushwood or netting, which are set athwart fish migration routes to guide the fish into enclosed pounds from which they cannot escape. The greatest of these are the *madragues* of the Mediterranean and similar constructions in Japan. These are formidable structures, which cost a great deal of money and employ many men. They are not set at random, but at sites used for many centuries and known to be productive; many of them were used by the Romans.

They consist of long lines of anchored nets running out seawards from the coast. These check the coastwise migrations of the tunas, and guide them into a square enclosure of netting, with a floor also of netting. The scenes in this 'chamber of death', where the thrashing tunas are gaffed out, are well known (see Fig. 85).

Prawns are frequently caught in filter-nets in estuaries; these nets are set out in long parallel lines and fish usually on the falling tide.

They may be just cone-shaped wicker work traps, from which the prawns are taken by a boat which moves up and down the line. Similar, but much larger, wicker work cones are used for the capture of salmon migrating upstream.

The stow-net is a more elaborate version (Fig. 90): it is a long tapering net of fine mesh, fished beneath a moored boat in a tideway, with the square mouth of the net held open by poles. Fishing begins when these poles are pushed down, and as the current sweeps fish and prawns into the net, the closed end is hauled aboard from time to time and the contents emptied out. The *ambai* of Thailand and Malaysia has the tapering net streamed from a scaffolding of poles driven deep into the seabed, and surmounted by a platform. Two long lines of stakes guide fish into the net, and boats moored to the scaffolding empty the closed end and sort the catch. Greatest of all are the *dōl* bag-nets of Bombay state, which are set from very long scarfed stakes set in water as deep as 29 metres (16 fathoms). These nets are 46 metres (150 feet) long, with a mouth of 76 metres (250 feet) circumference. Long bridles moored to massive concrete blocks support the stakes. The meshes gradually decrease as the net tapers, from 10 centimetres (4 inches) square at the mouth to 1·5 centimetres ($\frac{1}{2}$ inch) or less in the bag. The catch consists of prawns, shrimps and fish, including *Harpodon*, from which 'Bombay Duck' is made, and some huge Sciaenid fish which prey on the shoals of fish and prawns.

Gill-Nets and Tangle Nets

Gill-nets of large mesh are used for the capture of sharks and swordfish, which are valuable food fish (Fig. 81). Gill-nets are also used for the capture of salmon on the high seas, originally in the Baltic, but now off Greenland and Alaska and the north-east Pacific. Gill-nets catch fish because the fish may be held in the meshes by their gill-covers, and most fish cannot swim backwards powerfully. Other fish, and spiny lobsters, for example, are entangled in nets. Tangle nets are set on the ropes more loosely than gill-nets.

Before the war, gill-nets or drift nets were used on the greatest scale for herring, and hundreds of miles of these nets were shot nightly in the southern North Sea. French drifters of large size fished miles of drift nets for mackerel in the Atlantic to the south-west of

Cornwall. Nets of smaller size and mesh are used to catch pilchards, sprats and sardines. Gill-nets have the disadvantage that usually only one shot can be made per night, and success depends wholly on the skipper's judgement, which cannot be later corrected. But active nets such as ring-nets, purse seines and midwater trawls can be used many times per night, and so give the chance of locating and then staying on the best fishing. The change can be shown in the British statistics, for whereas before the last war most herrings were caught by gill-net or drift-net, in 1969 only 14 per cent were caught in this way, as compared with 31 per cent by the ring-net, 11 per cent by purse seine and 44 per cent by midwater trawl – all active fishing methods as compared with the passive gill-net. In some fisheries however the efficiency of the gill-net can be increased by beating fish into the net or in other ways, including the illegal use of explosives, causing the fish to panic and so gill themselves.

Encircling Nets and Midwater Trawls

When the ring-net or purse seine is used for the capture of surface fish, the shoal of fish is first located, and then the fishing boat encircles it at high speed, paying out the net as it goes. When the circle is complete and the fish enclosed, the foot of the net is either pulled to the surface by hauling it faster than the footrope, as in the ring-net; or else, as in the purse seine, the footrope is drawn together by a pursing line, thus closing the bottom of the net (Fig. 81). In both cases, the fish are landed by gradually hauling in the two ends of the net so that the fish become concentrated for bailing-out with large scoop nets. Power blocks are used to assist in the swift hauling of these nets and these, together with the use of sonar, are responsible for a great new herring and mackerel fishery in the North Sea. Purse seines of enormous size are used in the Pacific by American fishing vessels for the capture of tunas.

There are two kinds of midwater trawl. One kind is pulled along by two vessels, and is commonly used for the capture of small fish such as sprats. It is a long net of small mesh, with a square mouth. The two boats cruise about in company, each searching for dense fish shoals with its echo-sounder. When a strong echo is got, the two boats come together and pass a warp across. Then they both go ahead, paying out the net to the depth at which the fish are swimming. As much as two tons of sprats can be got in a few minutes, and this method has done much to revive the inshore fisheries.

The midwater trawl towed by one ship has had many forms, and its difficulty is that fish can escape either upwards or downwards. One commonly used type has the weighted footrope the same length as the floated headline. Towed at speed, and on the indication of the echo-sounder and sonar, it can achieve big catches. But on fish shoals swimming near the surface the purse-seine is the more reliable.

Drag-Nets

These are active nets which are dragged along to engulf the fish. Beach seines are in use the world over. They are large or very large nets, which are laid by a boat in a semi-circle from a beach and back to the beach, catching the fish enclosed in the semi-circle, when the net is hauled ashore. In the industrial countries they are tending to pass out of use except in very valuable fisheries such as the salmon, because they need a lot of hand labour.

In tropical countries where labour is readily available and cheap, seines are among the most effective nets. Beach seines worked on the Malabar coast of India have wings half a mile (800 metres) long; Indian migrants have carried the use of this net to Ceylon and Malaysia. Sometimes labour can be saved by using caterpillar tractors with power-driven winches to haul in the wings of the net. In earlier days, the pilchard of Cornwall was caught by seine; look-out men, called 'huers', watched the movements of the pilchard shoals from the cliffs, and signalled them to the teams operating the seine. But because of the problem of assembling the large teams of men, the pilchard fishery has now become a gill-net fishery.

The Snurrevåd, or Danish seine, is worked on the same principle as the beach seine, but is used in the open sea, and hauled back on board the fishing boat (Fig. 82). The warps attached to each end of the net can be a mile long, and the net itself may be very large, though light. It is a handy gear, and each haul can be done in an hour; so a large area can be searched for fish in a day's work. As with the ring-net, a buoy with one line attached is thrown overboard, and then the first warp is paid out as the boat steams at full speed. Before the net is reached, the boat turns through an angle, and after paying out the net it returns to pick up its buoy. Both warps are then hove in at speed on a special line hauler and coiler. Thus a large kite-shaped area is fished at each haul.

Because this gear can be used by a small and low-powered fishing vessel, it has to some extent displaced the small trawler. In the fisheries of the east coast of Australia, for example, since the war the Danish seine has superseded the trawl in the valuable flathead fishery.

Trawls

Probably the trawl is at present the most widely used and most productive of fishing gears, accounting for most of the world's catches of bottom-living or demersal fish (Fig. 80). It has also been modified to catch pelagic fish such as the herring, when these are living close to the sea bottom. The trawl is a big shaped bag-net, which can be towed at a fair speed along the sea bottom, and catches living fish within about 2 to 2·5 metres (6 to 8 feet) of the bottom. It is designed so that the headrope advances over the bottom ahead of the footrope, so that a fish cannot escape by jumping upwards. Thanks to the observations of frogmen and underwater television, much is now known about the action of the trawl. To hold the net open, there are two devices. One is the beam, a massive baulk of timber to which the headrope of the trawl is laced. The other is the pair of otter boards, which act as kites, forced apart by the water as they advance at an angle.

The heavy beam was difficult to handle in bad weather, and its weight set a limit to the possible width of the net opening. Hence the otter trawl, in which the mouth of the net could be very wide, ousted the beam trawl, especially when steam took over from sail. But the beam trawl has recently staged a revival. It has been found as effective as the otter trawl in the sole fishery, and the Dutch fleets recently put in 700,000 hours of fishing with these modernized beam trawls, while otter trawling declined by half in the same period. The way of working these trawls is the same as has been used in China for centuries. Instead of one trawl, the vessel tows two or more smaller trawls from booms. These can be hauled in turn without stopping the ship. In south China, prawns are the chief produce, and as many as seven trawls may be towed by one ship, from spreaders.

The otter trawl has been in use for centuries in the

Mediterranean, where otter boards of distinctive size and shape are used; but far older is the pair trawl, a trawl towed by two vessels, originally sail, but now powered, which, by keeping a certain distance apart, spread the net between them. This pair trawl has been evolved by the Spanish fishermen into a formidable fishing engine of great width, and this is the chief means by which they fish in the deep water of the Atlantic coast from Morocco north to Ireland. For a few years, as a result of the Spanish Civil War, a few pair trawlers started work at Milford Haven, in South Wales, and very successful they were. When two small steam trawlers towed a pair net between them, they caught more fish than a large and powerful trawler working singly.

Comparisons can be made between the performances of the different types of trawl. In the early days of power-fishing, steam trawlers using the beam trawl had an average annual catch of 196 tonnes (192 tons) of fish, sailing trawlers using the same gear 49 tonnes (48 tons). This superiority was of course mainly due to the independence of the steamer of wind and tide, giving it a far greater amount of fishing time. And when the otter trawl came into full use in the 1890s, the average catch of the ship using the otter trawl increased from 194 to 266 tonnes (190 to 261 tons) annually, this difference indicating the superior performance of the otter trawl over the beam trawl, when both were used by steam trawlers of about the same size.

In 1921, a modification of the otter trawl was introduced by Oscar Dahl, trawler owner of Boulogne. It was an adaptation of the pair trawl for use by a single vessel, employing otter boards to keep the net spread open, but using very long wire and combination bridles between the otter boards and the net. These bridles dragged along the bottom, and aqualung divers have confirmed that the dense clouds of soil stirred up by the bridles are effective in driving fish into the path of the advancing net. The bridles thus give the net a greater effective fishing width. But the original Vigneron-Dahl trawl was cumbersome to work, with its long bridles, much of them of combination wire awkward to handle, though its first results were sensational. These results were enhanced by the fact that the gear was introduced at the time of great post-war recovery of the fish stocks (p. 233); but there is no doubt that the wire bridles do enhance the catch, and have increased the fishing pressure on the stocks of fish.

Among steam trawlers using the same kind of trawl the performance, in terms of catch per 100 hours of fishing time, increases in simple and direct relation to the gross tonnage of the ship. And when diesel trawlers are compared with steam trawlers, the motor trawler is equivalent in fishing power to a steam trawler 50 gross tons larger.

Marine Fish Farming

All these different fishing methods are taking fish from a common resource, for the fish in the sea are the property of all. With the exception of the Pacific Halibut Treaty between Canada and the United States in 1935, it has so far generally proved impossible to get agreement among the fishing nations for a controlled exploitation of these resources. There is no agreement on exclusive fishing limits or on the fishing effort which can be permitted. The sad history of the baleen whales, and of many over-exploited fish stocks, show that where all can squander there is no sense in thrift. Therefore an attractive feature of the farming of fish is its exclusiveness; a fish farm is private property, so that thrifty husbandry by the farmer benefits him directly. He has an inducement to conserve his stocks of fish which is wanting to the owner of fishing vessels who shares the common resource with thousands of others.

Marine fish farming is already being done on a substantial scale in the brackish waters of estuaries and salt marshes. About one million acres are in production and, at an average rate of production of 400 pounds per acre per annum, contribute not less than 200,000 tons of fish and prawns annually. The stock must be salt-tolerant, well-priced, easy to feed, easy to breed (or at least with naturally spawned young which are easy and cheap to obtain), and fast-growing. The most usually cultivated are eels and grey mullet in temperate waters, and milk-fish (a large herring like fish), grey mullet, *Tilapia* and prawns in tropical waters. Shellfish also have for long been cultured, and their sedentary habit makes them a very suitable subject.

Various devices of husbandry enable the rates of production to be far higher than those of natural marine habitats. The latter are of the order of 4 to 30, and in exceptional cases up to 75, kilograms per hectare per annum: in marine fishponds production at a rate of 200 to 400 is common, with production up to 2,000 kilograms per hectare per annum in exceptional cases (Fig. 89).

Such high rates of production are got by measures hardly possible in the wild fisheries:

1 Carnivorous fish are excluded, and fish and prawns are raised which feed directly on algae and detritus, that is, near the level of primary production.
2 There can be population manipulation, so that there need be neither understocking, which would leave unused primary production, nor overstocking, which would waste resources by reducing the growth of the stock.
3 Only the desired fish need be grown, for unwanted predators and competitors, and harmful pests and parasites, can be controlled or eliminated. This can be done by periodical stocktaking and sorting, and the use of insecticides and pesticides to remove insects and snails which feed on the same algal food as the fish.
4 The natural fertility of the ponds is increased by manuring, mainly with organic fertilisers such as composts, which provide detritus as well as some nutrient salts. In the shallow waters of marine fishponds the inorganic fertilizers, such as nitrates and phosphates, seem ineffective. It has been suggested, however, that nutritious waste materials now dumped near the coasts might be disposed of in mid-ocean areas which are naturally poor in nutrients (see p. 209), and so enable them to support a richer marine life.

The very high rate of production in mussel beds is due to these molluscs feeding direct on the phytoplankton, thus at the level of primary production. By growing mussels on ropes suspended from rafts, so that they can feed on the whole water-column, and be raised clear of the usual predators, the enormous rate of production of about 300,000 kilograms per hectare per annum is feasible (Fig. 86). The other fish grown in marine fishponds, milk-fish and grey mullet, feed on the algal pasture growing on the bed of these shallow ponds. The rate of production of this algal pasture may be as high as 28,250 kilograms per hectare, which supports a fish crop of 2,500 kilograms per hectare. The prawns grown in the ponds are partly herbivorous and partly carnivorous on smaller crustacea.

There is great scope for expansion of this well-established kind of fish farming; for it has been estimated that, in addition to the existing million acres of ponds (themselves capable of much increased rates of production), nearly 2 million acres are available for development as fish farms in the Niger Delta of West Africa, 6 million acres in Indonesia, and half a million in the Philippines. There must also be many millions of acres of potential fish farms in the deltas of South America.

It has already been shown that the deeper oceanic waters contain, below the lighted zone, rich resources of plant nutrients; and that, when upwelling brings them up into the lighted zone, they support vast crops of phytoplankton, which in turn support the world's greatest fisheries. Now there are plans to pump this deep water up into the confined waters of oceanic atolls, the ring-shaped coral islands surrounding a shallow lagoon, which so much interested the young Charles Darwin. These atolls are built on the emerging or submerging tops of steep submarine mountains (see pp. 77–8), so that a pipe of no very great length could draw water from several hundred feet down, water far richer in salts such as phosphates and nitrates than the surface water (p. 209). Trials at St Croix, in the American West Indies, have shown that plant production in sunlight in this rich water may exceed 200 gmC/m^3/hour, or say 2 gmC/m^3/day. This is as high as the rate in the Humboldt Current (Peru Current) on which the world's richest fishery is nourished. It has been calculated that an atoll lagoon, continuously replenished with this pumped deep water, and stocked with edible seafood able to make full use of the plant material, obviously mainly phytoplankton, could produce enough protein per square kilometre to supply the needs of 4,600 people continuously (see also p. 201).

Less interesting from the production aspect is the technique, developed in Japan but now being copied in both Europe and America, of rearing fish and prawns very intensively in cages staked to the sea bottom, or floated from rafts. Very high rates of production are recorded; but this is not new fish production, but production got by feeding fish to obtain fish, since trash fish or pelleted foods based on fish-meal are the chief foods given in these cage-cultures. All that is gained is an increase in value, from cheap trash fish or fish pellets to valuable fish such as the yellowtail, in Japan, salmon and rainbow trout, and possibly in the future soles and turbot in Europe.

C.F.H.

14 CONSERVATION OF THE RESOURCES OF THE SEA

Conservation is defined in the Shorter Oxford English Dictionary as 'the act of conserving; preservation from destructive influences, decay, or waste; preservation in being, health, etc.' For mankind, conservation of the sea's resources means their management so as to obtain the maximum sustained yield of useful material.

In 'Basic Food Resources of the Sea' (p. 210), it was stated that the annual production in the sea, reckoned as carbon, is about 16,000 million metric tons, but it is not known if this huge total fluctuates from year to year. There is some evidence that plankton production may have declined over the last two decades. We make but poor use of this bounty, only about ·038 per cent, almost wholly in the form of fish.

The total production of the land is about the same; yet how much better is the use we make of the production of the land. Food grains such as wheat and rice; roots and tubers such as potatoes and turnips and cassava; grasses which we eat at one remove as mutton, beef and veal, or as milk and all its products: all these are the direct use of plant production. About 80 million metric tons of dressed carcass meat are produced annually. Timber, again a direct use, gives enormous quantities of material for technical use.

But we use none at all of the plant production of the sea unless we include the few thousand tons of seaweeds collected, or artifically grown, as a foodstuff or raw material for industry, and most of the fish we eat are predators for a part or the whole of their food. It was shown in Chapter 12 that predators, being at the bottom of the food chain, yield little of the original primary production. In sum, from the land we take primary plant production, and herbivorous animals only one stage from it; while from the sea we take largely carnivores, well down the food chain. This seems to be the negation of conservation.

Limits of Increased Production

Some very different estimates have been made as to the possible increases of fish production from the oceans. While it is generally agreed that there could be a marked increase in the harvest of fish, estimates of this increase vary from 4 times to 10 times and more. One estimate places possible fish production at 2,000 million tons, as compared with the present 50 million. But recent estimates have not been so optimistic. There is some reason to believe that the production of fish from the present fishing grounds is nearing the limits of what can be got from them on a permanent basis, given the present rate of primary plant production. In the well-fished North Sea, it is considered that the catches of fish, which have been increasing fast since 1960 because of some technical improvements and the occurrence of some successful breeding years, are up to, or even in excess of, the total potential fish production.

This may not be so for other areas; for instance, it is believed that there are great unexploited resources of anchovy and hake in the upwelling waters of the California Current, which might yield a million tons annually. But those who think that even the present rate of fish harvesting is nearing what the sea can produce are looking at the possibility of catching for food, not the zooplankton-eating fish (though these are about nine times as economical converters of primary production as the predators), but the zooplankton itself. Many people have tried eating zooplankton; it has been described as having a shrimp-like flavour, and of course it would be as nutritious as any other marine crustacea. A major promotional campaign would have to be mounted to make a public demand for a new product like zooplankton, and an acceptable way of selling it. But it is time to start trials.*

* There is a local coastal fishery for Euphausiids in Japan; and in Malaysia the pelagic mysid *Acetes* swarms in the estuaries and is caught and processed into a much-relished paste.

A Zooplankton Fishery?

A practical problem in fishing for plankton is that it is thinly dispersed in three dimensions, over many fathoms in depth. The difficulty can be illustrated by some calculations on the cost of catching plankton. Even under favourable conditions, the cost of a ton of dried plankton would be from 6 to 10 times that of the present commercially caught species of fish.

But we have yet to develop a technique for hunting plankton, which is very patchy in distribution. At present, we shoot our plankton nets on speculation, whereas we shoot our fishing nets on well-known indications, and if these are wanting we do not fish. But at least one major expedition is out exploring the possibilities. The Russian research ship *Akademik Knipovitch* (Fig. 65) is using two plankton nets of 20 metres (66 feet) width each, and is no doubt gaining experience with sensitive echo-sounders and asdic to locate plankton concentrations. This expedition has found zooplankton in commercially interesting quantities, but not consistently enough for a viable fishery. The greatest resources of zooplankton are the swarms of 'krill', the oceanic shrimp *Euphausia superba* (Figs 60/61), in the Antarctic Ocean, which is the chief food of the baleen whales and of the crab-eater seals and penguins. It has been calculated that before 1930 there were more than 400,000 blue, fin and humpback whales in the Antarctic, and that they must have consumed over 77·3 million tons of krill each year. Now that these whales have been decimated, there must be a vast stock of these shrimps available to feed us if we could catch them economically. The great whales certainly had the means to detect swarms of krill, as some recent work has confirmed, and we also must learn to do so.

Plankton-Feeding Whales and Fishes

In the meantime, we must still depend on those creatures which feed on plankton; it has already been said what efficient converters of phytoplankton to flesh are the mussel and oyster. *Euphausia* itself is a direct feeder on phytoplankton and has fine filters of bristles on its thoracic limbs. By feeding on this shrimp, the whales take the primary production of the sea at only one remove.

Though most of the fish we eat are large, the plankton-feeding fish are often small, such as sardines, sprats and anchovies; much smaller are the bristlemouths and Myctophids, very small fish, probably the most abundant of all, which represent the greatest unexploited resource of the sea. But we still have to learn to catch them economically, and even then, they would have to be processed to fish meal.

Unexploited Fish Resources

The blue whiting *Gadus poutassou* feeds on Euphausiids as well as on Myctophids, and is itself the chief food of the hake in the deep water to the west of Europe. Blue whiting represent a resource estimated to be of the order of several hundred thousand tons, ready for exploitation, and a Norwegian research vessel, the *G.O.Sars*, is searching for them. Caught direct for human food or for processed food, blue whiting would give about nine times as much food as the hake which feeds on it.

There are probably many other potential fisheries not at present exploited because they would not pay at present prices. A rise in the price of fish, relative to the cost of catching them, might open up some of these possible fisheries, and so use more of the resources of the sea. But most of the future additional fish supplies, whether of plankton, small oceanic fish, or industrial fish not intended for human consumption, would have to be processed cheaply into a form which could be easily stored and transported, and sold to consumers at a price they can afford. The anchoveta of Peru, of which 9 million metric tons were caught in one year, are bid for at prices far beyond the means of peasant people, to be processed into high-class fish meal, and sold overseas to the livestock industries of the industrial countries. The world's mass-production of pork, veal, chickens and eggs has come to depend to a high degree on this supply of fishmeal. This is also bad conservation, for most of the protein in the fishmeal is lost in conversion into poultry and livestock protein.

Yet attempts to market a high-grade fish meal for human consumption have so far failed, such is the conservatism of our food-habits. But there is some hope of making it acceptable to children by incorporating it in cereals, to make high-protein biscuits and the like.

Scrap fish and shrimp are already processed on a small scale, especially in some countries of South-East Asia. They are fermented to make fish sauces and pastes. Small anchovies, other scrap fish, prawn and shrimp trash, and other material is pounded in mortars with salt, and then placed in big earthenware pots to ferment for weeks or months. If a sauce is to be made, fermentation goes on until a rich brown liquor rises to the top of the pounded mixture, and this is skimmed off and bottled as first-grade sauce. There may be a second fermentation to give second-grade sauce. This fish sauce, called by so many different names, is a deep brown liquor, with a strong salty flavour like that of beef-extract, and so rich in protein that it will coagulate in a warm spoon.

Alternatively, the raw materials, as above, may be pounded into a stiff paste with cassava or tapioca, and fermented. When fermentation is complete, the paste is pressed into round red puddings very much like Dutch cheeses in appearance. Both the sauce and the paste are strongly flavoured, and help in the mastication of flavourless starchy foods. They keep well, and their high protein content gives a well-balanced diet with a starchy food.

Discarded Fish

The world's supply of fish might be increased if all the material caught by the fishermen were used. Most fishing methods produce a proportion of unwanted fish and invertebrates, which are either unmarketable or so low-priced that they are not worth carrying home. This is still true, though over the decades more and more species formerly discarded are now brought home. In the 1880s little except the prime fish, turbot, brill and soles, was worth bringing to market, most of the rest of the catch being discarded. Nowadays most of the catch is brought home. The rejected material consists of unsaleable and damaged fish, fish guts, and invertebrates such as starfish, shellfish, sea-cucumbers, sponges, soft corals, and squids and other molluscs, though recently a sale for squids has grown as substitutes for abalone in Chinese restaurants. Formerly there was a sale for them at low prices as bait.

The advent of the factory trawler is remedying this. The factory-ship has long been in use in the whaling industry, and is now in use in the high seas fisheries. In these ships, all material caught is used, the best fish for filleting, and the rest for fish meal and oil. This is good conservation.

Deep-Sea Fish

Another possibility is to extend the fisheries into deeper water. Well before the last war, trawlers from the west coast of the British Isles were working in water as deep as 600 metres (2,100 feet), and long-lining in the Pacific for halibut may be done as deep. I have seen experimental commercial trawl hauls made in water as deep as 900 to 1,000 metres (3,000 to 3,600 feet). The catches were not big, and consisted mainly of deep-sea fish such as the grenadiers, deep-sea lings and *Mora*, a cod-like fish. We tried fried fillets from these fish, and though they tasted no different from more familiar fish fillets, their big heads and long thin bodies did not give much edible meat. It seems doubtful if any great new resources are to be found there, though there is one exception. The Russians have discovered a payable fishery in depths of 600 to 900 metres (2,100 to 3,000 feet) off Labrador and Baffin Land. The catches are mainly the grenadier fishes and the halibut, and the season is from June to December. The hake off the west coast of North America is a deep-sea fish, and the blue whiting, mentioned earlier, lives over very deep water, and would have to be caught with some kind of midwater trawl.

Overfishing

The existing stocks of fish could give better crops if they were managed internationally. The blue whale, probably the largest animal that has ever existed, in the heyday of its fishery provided millions of tons of oil, meat, fertilizer and pharmaceuticals, converting with efficiency the animal plankton, and especially the *Euphausia*.

In 1930, about 30,000 blue whales were killed; but since that year the catches of this splendid animal have declined until they were almost extinct by 1963; it is now a protected species. In 1931, because of the lack of demand due to the economic depression, there was a voluntary limitation on whale hunting; so management could have been carried out if there had been a will to agree on this. Especially disturbing was the failure of the whale stocks to show a recovery after six years of respite from persecution during the war years 1940–6.

Modern fishing takes a high proportion of the fish stocks, but luckily they have the power to make a rapid recovery when given a rest, as during world wars.

Fish populations have great natural fluctuations which are independent of man's predation. In the late Middle Ages, the great Hansa Federation rose to wealth and power on the proceeds of a rich herring fishery in the southern Baltic; in fact the herring was adopted as the badge of the Hansa. This fishery disappeared two centuries later. A winter herring fishery at Plymouth, England, died out in the 1930s. In both these cases, there was a change in the hydrographic conditions which prohibited the continuance of the herring population. But it is very interesting to know that the hydrographic conditions in the western English Channel are changing again towards those prevailing formerly, and that with this change there are signs of the reappearance of a herring fishery.

A great fishery for pilchards in California took as much as half a million tons in 1930, but it vanished and has been replaced by an anchovy fishery. The oil-sardine of the Malabar Coast of India supports a fishery which may fail altogether in some years, giving rise to widespread distress, and in others may give catches too large to handle. The great increase in the Arctic cod fishery in the 1930s was associated with a slight warming-up of the water; should the hydrographic conditions change to give another cold period, the European fisheries would suffer. Such changes are due to causes outside man's control.

But man can now affect the size of the stocks of fish, and thence of the harvest he gets from them. In the case of several fisheries it has been proved that the catch of fish per ship varies with the number of ships at work, or with the fishing effort employed: this could happen only if a large proportion of the fish stocks were being caught.

The overfishing problem has been debated for several decades, and it has long been known that a reduction of the fishing effort would, in an overfished fishery, result not in reduced catches, but in increased sustained catches of better quality fish, and obviously with a reduced fishing cost. The first application of this was in the halibut fishery of the North Pacific, a fishery shared by fishermen of Canada and the United States. From the beginning of commercial fishery in

1888, catches rose to a maximum of 30 million kilograms (70 million pounds) in 1915. But these high volumes were only maintained by opening up new and more distant grounds as the nearer ones were worked out, and by setting ever more fishing gear. An International Halibut Commission was set up to research on remedies and, taking advantage of the depression years of 1930–1, when catches were much reduced because people could not afford to buy halibut, the Commission set a limit of 20 million kilograms (46 million pounds) for the next seasons' catches, and the fishery was declared closed when this total had been caught. The stocks of halibut began to build up again, and in later years the permitted offtake was progressively increased, since they were now a smaller proportion of a larger stock. On the worst affected southern grounds, the abundance of halibut increased by 60 per cent. By easing the rate of fishing, the life-expectation of the fish was increased, and more fish survived to a large size and weight.

The effect of the two world wars showed this equally well. The result of a much reduced fishing effort in the North Sea in World War I was, by 1919, a 100 per cent increase in the fish stocks. World War II resulted in a 400 per cent increase. On the British south-west fishing grounds worked from South Wales, a 60 per cent reduction in fishing effort resulted in a six-fold increase in the catches per vessel, and 70 per cent of this improvement was due to the increase in the number and weight of large fish, the result of a reduced fishing mortality.

In the important flathead fishery of New South Wales, the respite due to the war-time requisitioning of the fishing fleet resulted in a great increase in the abundance of these fish.

These, and other examples, show that a planned reduction in the fishing effort would result not in smaller but in larger catches, got at less cost. But such management of the fishing effort would have to be agreed internationally, and seems scarcely possible. Fishing industries provide home-produced food, employment for thousands of fishermen who are also potential naval reservists, and employment for thousands of people in the ancillary industries. In many cases, a fishing industry earns foreign exchange for the country. The tendency is rather to encourage and even to subsidize a big fishing fleet.

Transplantations

Where there are supplies of unused natural fish foods, a good conservation measure is to transplant suitable fish to the locality. Thus plaice from the overcrowded nursery grounds off the west coast of Jutland are transplanted to the inner Liimfjord, a sheltered waterway which almost divides the Jutland peninsula. There they grow well, and provide a fishery wholly within Danish waters.

Juvenile plaice from the same overcrowded nursery grounds grow very fast when transplanted to the Dogger Bank, putting on weight (and cash value). Since the early days of this century, the suggestion has been canvassed that this should be done, and in the late 1930s a tender was actually submitted to the International Council for the Exploration of the Sea. It seemed very reasonable; yet nothing has been done. The Dogger Bank is in international waters; nobody has solved the problem as to who pays for the transplantation when all are free to catch the fish. Even if a future United Europe were to carry out this sensible scheme, it would still be open for non-member countries to poach the fish perfectly legally.

The Russians have successfully transplanted a Pacific salmon to their European waters. This species has the advantage of returning to the rivers at an early age, and at a size and shape very suitable for canning.

Salmon and sturgeon fisheries are now maintained in some cases by artificial breeding (Figs 87 and 88). The ripe fish are caught and stripped of their eggs and milt. The fertilized eggs are hatched, and the young fish raised in tanks on special foods until they are large enough to be released. Though such 'hand-reared' fish might be expected to be fair game for enemies in a rough world, experience in Sweden is that these fish in fact show a slightly better survival rate than wild-spawned fish.

Industrial Fishing

Industrial fishing, which catches fish as a raw material for the feedstuffs industry, wastes a great deal of potential food, and so is anti-conservationist. Since the feeding of fish meal to livestock may lose up to 90 per cent of the protein present (see p. 210), a man from Mars would wonder why the fish meal was not fed directly to humans.

In the eastern and north-central North Sea, an industrial fishery for baby one-year herrings began in 1945. By 1951, this fishery had grown so fast that over 800 million baby herrings were caught annually. But, simultaneously, the East Anglian autumn herring fishery, then the greatest in the world, died away. The baby herrings were rendered down to fish meal, chiefly for the Danish dairy and bacon industry. As the late Dr Hodgson, eminent for his work on the herring of the North Sea, put it, 'the herring stock of the North Sea has been capable of feeding a large part of Europe's human population for hundreds of years; but it has yet to be proved that it can, in addition, feed Europe's livestock.' The proof is still wanting.

Pollution of the Sea

As the population of the world increases, its harmful waste-products have created a world-wide disposal problem. Human rubbish such as flashlight batteries and empty cans have been dredged up from the bottoms of the deepest oceanic trenches. Fish have been found with their intestines filled with small beads of polystyrene originating in manufacturing industry. In earlier days, the discharge into the sea of human, animal and industrial wastes was relatively so small that the oceans could purify them. Even so, however, there were local problems, where raw sewage was deposited on beaches; and recently there have been outcries about the pollution of the bathing beaches north and south of Rome by the raw sewage of the city. Domestic sewage also contains much phosphate from detergents, so that it is a rich nutrient solution which can support dense growths of phytoplankton. These in turn, dying down, produce a mass of organic matter, which adds to that in the sewage to deoxygenate the water, and especially the deeper water. Many areas of the deeper parts of the Baltic Sea are now devoid of oxygen.

Pollution of this kind by wastes discharged from the coasts would be lessened by taking and dumping them far out at sea; in fact, if discharged into barren areas of the oceans, they might indirectly increase the world's supply of fish. The knowledge of surface and deep ocean currents that we now have makes harmless or beneficial dumping of wastes an easier possibility (see pp. 33 and 34). But thus disposing of wastes would be expensive, and the cost would be passed on to the consumer who, in the end, has to decide how much he is ready to pay for a stand against increasing pollution.

Industrial wastes include heavy metals and dredging and mining spoils. China clay wastes are a local problem; many people in Japan have been poisoned by toxic factory wastes discharged in the Inland Sea, and it has recently been shown that excessive amounts of heavy metals in solution can inhibit the development of larval stages of marine organisms. Dangerous wastes such as radioactive materials are usually dumped in the deep ocean in steel or concrete containers, and so also are dangerous chemical warfare wastes.

A new danger comes from the new and very powerful pesticides, chlorinated hydrocarbons. These are often sprayed on crops, and the spray may be carried up and distributed by the wind over wide areas. Residues of DDT have been found in the fat of Antarctic penguins; and it is estimated that one billion pounds of DDT are distributed about the biosphere. They are only slowly broken down, and are accumulated in marine animals by the food chains. Though not at present in lethal quantities in sea-water, they become concentrated by thousands of times in the food chain from plant plankton to predators, and DDT and the other chlorinated hydrocarbons have affected the breeding of some seabirds (see Fig. 79), and possibly of crabs. Their potential for harm will not be known until more years have passed; but their presence in the sea is plainly harmful. The dilemma is serious, for much of the increased food production which has recently kept pace with the increase of the human population is due to the use of these pesticides.

The modern world runs on oil, and vast quantities are transported across the oceans. A common source of oil pollution in the past was the deliberate discharge of small quantities of oil in the course of tank cleaning operations. National and international regulations, and the introduction by the industry of expedients such as the 'load-on-top' method, have lessened this. In this method the mixture of sea-water and oil that results from washing out tanks is kept in the tank rather than discharged into the sea, and the new cargo loaded on top of it. The dangers of accidental spillage resulting from collisions at sea are however growing with the ever-increasing size of individual oil tankers. The *Torrey Canyon* disaster was only one well-publicized example. When oil is spilled into the sea, a serious nuisance is created. Some marine life, especially birds (Fig. 73), is harmed directly, and great harm is done to the recreational potential of beaches when the oil drifts ashore. Even here, however, there is continued biological purification, for shellfish such as limpets, chitons and crabs browse on the oil film on rocks, and pass the oil out as pellets which are easily dispersed (Figs 76 and 77).

Left to itself, an oil spill in the open sea slowly disperses, partly by the evaporation of the lighter fractions of the oil, and partly by the slower oxidation of the oil, chiefly by bacteria. But the heaviest fractions drift about and have been reported in large quantities in mid-ocean.

There are now organized teams and equipment for dealing promptly with oil spills in many maritime countries. The oil slicks may be sprayed with dispersants (Fig. 75), which break them up into small droplets which are more quickly attacked by bacteria and will often sink. An alternative is to spray with treated sand (Fig. 74) or chalk, which sinks the oil. But this may not be the end of the damage done, for the sunk oil will continue to oxidize. It is estimated that one litre of oil will use up all the dissolved oxygen in 400,000 litres (88,000 gallons) of sea-water. Especially in deep and very deep water, replenishing the supply of oxygen once it is used up, would be very slow, and oil successfully dispersed from the surface may create lifeless areas on the sea bottom.

C.F.H.

15 WHO OWNS THE OCEANS AND SEABED?

One of the great political questions confronting the United Nations is concerned with the question of ownership of the oceans and the seabed beneath them. An early attempt to divide the oceans arbitrarily was the papal bull of Alexander VI in 1493, which declared that discoveries made 100 leagues west of the Azores and for a space of 180° longitude should belong to the Portuguese, while those to the westward of the same meridian and for the same space (180° longitude) should belong to the Spaniards. It is of interest that neither party adhered to the agreement, and one excuse was that longitude was very difficult to determine accurately, particularly at the borderline of the zones.

In the early 1920s the legal adviser to the British Foreign Office wrote an article in which he posed the question: Whose is the bed of the sea? The article drew four conclusions about the international rights of any state concerning the water round its coasts and the seabed beneath it. These conclusions received wide international acceptance and reiterated the idea of seabed ownership extending three miles seawards of low water mark. It was foreseen that a nation could extend its limits if it constructed a tunnel outwards which had been started inside its territorial waters, but it was not foreseen that in less than a quarter of a century nations would begin to be very much interested in claiming the entire continental shelf adjacent to their national shorelines.

In 1945, the United States made its now famous proclamation that claimed all the natural resources of the subsoil and seabed of the continental shelf contiguous to the coasts of the United States. This was interpreted as a zone extending to a depth of 200 metres (660 feet). Thus in effect the United States laid claim in some instances to the seabed 400 kilometres (250 miles) from its coasts. This began a whole movement, and soon after other states forwarded similar claims to their adjacent waters. For example, Brazil announced that its jurisdiction stretched to 480 kilometres (300 miles), above and below the surface. Peru and Chile claimed 200 miles. Iceland and Canada claim an 80-kilometre (50-mile) non-fishing limit, and a 160-kilometre (100-mile) non-pollution zone.

An attempt was made in 1958 to obtain agreement within the framework of the United Nations. At this time the Assembly considered the draft proposal which the International Law Commission had drawn up in 1956. One of the proposals put before the U.N. was officially to define the continental shelf at a depth of 200 metres (660 feet). This was counteracted with a British and Dutch counter-proposal to extend this depth to 550 metres (1,650 feet), but this was not agreed, and the 200-metre lobby won the day. The U.N. adopted four Conventions which were ratified by many members, although not by all. The Conventions cover the freedom of the high seas – affecting nations both with and without coastal waters, freedom of fishing and navigation, the rights to construct pipelines and submarine cables on the seabed, air passage over the oceans, and in addition all the accepted freedoms associated with the general principles of international law.

Since the 1958 Convention, the 200-metre depth limits of continental shelf water appear to be operating reasonably well. However, one of the principal disputes was the definition of the width of national rights on the shelf and in the open sea. In 1967, Malta proposed that the oceans outside the shelf should be declared a common heritage of all mankind, and at present thirty-five member nations form a U.N. committee to study the subject.

One tenative idea is to divide all the oceans of the world into separate lots to ensure fair shares for all nations under U.N. control. In this idea the oceans would be divided into 30,000 blocks measuring up to 20,800 square kilometres (8,000 square miles) each. If a particular government did not wish to use its allocation, it would be able to sublet it to others. This subdivision has worked well in the case of the North Sea bottom which has been divided off between member nations whose shorelines border it. Nevertheless, if this latter system were applied to the oceans as a whole, then nations with long coastlines would gain the lion's share, and nations without shorelines nothing at all. Whatever scheme is finally adopted, be it free-for-all or by allocation, it will not come about without much debate, conflict and controversy. In the future it is likely that the seas and oceans will become the international political arenas where opposing ideologies clash head on.

P.L.B.

INDEX

Figure numbers are printed in bold type. Page numbers in italics indicate the main reference on a subject.

Abalone 230
Abidjan 201
Abu Dhabi 195, 197, **71**
Abyssal plain *14*
Afar Triangle 62
Africa, South 175, 209, **56**
African Rift Valley 19
Agassiz, A. 172
Air currents, definition of 23
Airy, Sir George 44
Alaska 218
Albert of Monaco, Prince 172
Aldabra Atoll 81
Aleutian Trench 18
Algae 224
Alkalinity 206
Ambai 218
America, North 225
America, South 225, **57**
Anchoveta 229
Anchovies 211, 215, 227, 229, **64**
Ancillary industries 233
Anemone, sea 212
Annual cycle, temperature 208
  sunlight 208, **58**
Antarctic Convergence 22, 208
Antarctic Ocean 7, 9, 12, *21–2*, 28, 175, 208, 228
Aqualung 179, 222
Arctic Ocean 7, 11, *20–1*, 174, 208, 232
Aristotle 1–2, 167
Arizona, University of 197
Artemia 204
Artificial breeding 234
Artificial freezing 96–7, **71**
Artificial reefs 215
Asdic *see* Sonar
Atkins, W. R. G. 206
Atlantic Ocean 9, 11, 16, *18–19*
Atlantis (legend), reference to supposed location 74–5
Atlantis II Deep 62
Australia 168, 221

Bacon, Francis 49
Bacteria 203, 212, 237
'Bacterioplankton' 203
Baffin Bay 170
Baffin Land 231
Bait 216, 230
  tanks 216
Baltic Sea 12, 19, 218, 232, 235
Banks, Joseph 169
Barents Sea 208
Barton, Otis 178
Basking shark 211
Bathypelagic Zone 71–3, **35**
Bathysphere 178
Beach seine 220
Beam trawl 221–2
Beaufort Scale, table 43
Beaufort Sea 29

Beebe, William 178
Benthic zone 71–3, 170–1, 203
animal life in **35**, **62**
Bering Sea 12
Bering Strait 27
Bermuda 174
Biological purification 237, **76**, **77**
Bioluminescence 72
Blake Plateau **23**
Blue-green algae 203
Blue whale 231, **61**
Blue whiting 229, **64**
Boron 8, 202
Bottom-living *see* Demersal
Boulogne 222
Bright lights 215
Brill 230
Bristle worms 215
Bristlemouths 213, **64**
Britain 167
Brittle star 212
Bromine 8, 202
Brooks, C. E. P. 34
Byzantium 168

Cage culture 225
Calcium 8, 202
Canada 223
Canning industry 234
Carbohydrate 205–7
Carbon 8, 207, 210
Carbon dioxide (CO_2) 205–6, 212
Carbon-14 dating 13–14, 33, 206
Caribbean Sea 9
Carlsberg Ridge 19, 58, **30**
Carpenter, W. B. 171
Caspian Sea 9
Ceylon 220
Chalk 237
Challenger Expedition ix, 18, 53, 77, 184, *passim*
Challenger Rise 18
Chamber of Death **85**
Chemical warfare wastes 236
China 168, 221
China clay 236
Chiton 237, **76**
Chlorinated hydrocarbons, as pollutants 236
Chlorine 8, 202
Chlorophyll 204–5
Chronometer, Harrison's 169
Chumming (ground baiting) 215
Cladocera 203
Claude thermal turbine 201, **69**
Cobalt 8, 202
Coccoliths 203
Cod 167, 211, 232
Coelacanth 69
Columbus, Christopher 167–8
Conservation 226 *et seq.*
Continental borderline *see* Continental shelf
Continental drift *48–64*, **29**
Continental shelf *12–14*, 22, 64, 209, 212–13
Continental slope 12, *14*, 231
Cook, Captain James, R.N. 169, 195, **17**
Copepods 203–4, 208, 210
Coral reefs 13, *75–81*, 170, 201, 203, 225, **36**, **38**
ages of 79
origin of 76–8
types of 77–8
Coring 175
Coriolis Effect (Force) 24, *25–6*
Coriolis, Gaspard 25
Cornwall 198, 219–20
Cousteau, Jacques Yves 179–80
Crabs 216, 236–7, **91**
Crawfish 216, **91**
Crown of Thorns (starfish) 79–80, **37**

Cuba 201
Currents *see* Ocean currents
Cuvier, Georges 48

DDT 236, **79**
Dahl, Oscar 222
Dampier, William 168
Danish seine 220, **82**
Darwin, Charles 75–8, 169–70, 225, **18**
Deep Scattering Layers (DSLs) 69–71, 213
Deep-sea deposits (sediments) 56, *64–5*, 175, **20**
Deep Submergence System Project (DSSP) 179
Deeps *see* Trenches
Demersal fish and fauna 204, 212, 221, **64**
Denmark 174, 210
Depths of the Ocean 173
Desalination 194 *et seq.*, **71**
Detritus *see* Organic matter in ocean
Devil fish 211
Diatoms 203–4, 206
Discovery Deep 62
Discovery expeditions 175
Dispersants 237
Dissolved salts 202
Distillation 195
Dogger Bank 234
Dōl nets 218
Dolphins 73–4, 213
Donn, W. L. 37
Downwelling 209
Drag net *see* Beach seine
Drake, Francis 168
Dredges 170–1, **79**
Drift nets 218, 220, **81**
Drifts, definition of ocean 26
Dunwich 168

Earth, the; age of 3, 49
gravitational constant (G) of 54
rotation period of 80
theory about construction of 54
theory about expansion of 54–5
Earth's magnetic field 53, 56–8
polarity changes in 53, 56–8
East India Company 168
Echo-sounding 214, 219–20, 228
Eddies 28, 209
Eel 174, 212, 223
Egypt 168
Eilat 197
Energy 205
Eniwetok Atoll 77–8
Enzymes 205
Equator 210
Euphausia superba 68, 228, 231, **61**
Euphausiids 210, 213, 229
Euphotic zone 67–8, **35**
Europe 164, 225
Ewing-Donn Theory 37
Ewing, M. 37

Factory ships 230
Faecal pellets 212
Falkland Islands 171
Faroe Islands 171
Feeding *see* Food web (food chains)
Fermentation 229–30
Ferrel, William 25
Ferrel's Law 25
Fertility *see* Production
Filter feeders 203, 205, 212
Fish 202, 212, 214, 218, 225
Fish culture *see* Fish farming
Fish farming 195, 201, 214, 223, **89**
production 223–4
Fish, fences 217
meal 226, 229, 234
ponds 224
sauce 229

Fish, shelters, artificial 215
stocks 231–3
Fisheries 210, 214, 219, 225, 231
Fishing, effort 223, 232–3
limits 223, 231
pressure 223
Fitzroy, Captain 170
Fixed fishing gears 216–17, **90**
Flash boilers 196, **71**
Flathead 221, 223
Fluctuations, natural 226, 232
Food web (food chains) 172, 203, 208, 210–11, 227, 236, **35**, **60**, **63**
Forbes, Edward 170
Fossil, corals 80
Fossil seas **21**
Fossils 4, 50
Franklin, Benjamin 30, 170, **19**
Freezing 195–6, **71**
Frogmen 221
Fruits 204

Gama, Vasco da 168
Gas *see* Oil and gas
Geiger counter 207
Geological time scale, table 5
Gerstner, Franz 44
Gibraltar 171
Gill nets *see* Drift nets
Gold, in the sea 183–4, 203
Gondwanaland 50
Grabs 170, **79**
Grains 226
Grasses 226
Great Barrier Reef 76–9
Greenhouse effect 36
Greenland 208, 218
Grenadier fishes 231, **62**
Grey mullets 223–4
Growth 206, 210
promoting substances 204
Guernsey 195
Gulf of Mexico 9
Gulf Stream *see* Ocean currents
Guyana 206
Guyot, Andre 16
Guyot (seamount) 16
Gyres (gyrots) 26, 206

Haber, Fritz 174
Habitats, manned undersea 182–3, **40**, **42**, **43**
Hake 227, 229, **64**
Hakluyt, Richard 30
Halibut 169, 216, 223, 231–2
Hamburg 167
Handlines 216
Hansa 232
Harpodon 218
Harrison, George 169
Hawaii 201
Hawkins, Richard 168
Heat 204, 208
Heavy metals, as poisons 236
Herbivores 210, 225
Herring 211, 214, 218–19, 221, 232, 235
Hess, Harry H. 16
Heyerdahl, Thor ix
Hjort, Johan 173
Holland 167, 221
Hopkins marine station (California) 173
Horse latitudes, origin of term 28
Humboldt Current *see* Ocean currents, Peru
Humboldt, F. H. A. 49, 76
Hutton, James 48
Hydrography 175
Hydroponics 197
Hydrosphere, definition of 1
origin of 1–6

Ice Ages 7, 13, 21, 35–40, **21**
possible causes of 36–40
Iceland 167, 208
India 168, 220, 232
Indian Ocean 9, 11–12, *19–20*
Indonesia 225
Industrial fishing 234
Industrial wastes 235–6
Insecticides 224
International Council for the Exploration of the Sea (ICES) 173, 234
International Geophysical Year (IGY) 175
International Indian Ocean Expedition (IIOE) 176
Ipswich 197
Iron 8, 202
Islands 74–5
approximate number of, in the Pacific Ocean 75
environments of 81
origin of, due to sea-floor spread 58, 75

Japan 216–17, 225, 236
Japan Current *see* Ocean currents, Kuroshio
Jutland 234

Kermadec Trench *see* Tonga-Kermadec Trench
Key West 195
Krakatoa 46
Krill *see Euphausia*
Kurile Trench 18
Kuwait 195
Kwajalein Atoll 201

Labrador 231
Labrador Sea 9
Lamellibranch molluscs 212
Lancaster, Sir James 168
Land, food production on x, 226
Lantern fish 68, 70–1, 213, 229
Laurasia 50
Life in the deep sea 170–1, 175, **62, 63**
Life support systems *see* Undersea life support systems
Lim Fjord 234
Limits to production 227
Limpets 237
Littoral zone 42, 64
Lobsters 216, 218, **91**
Lofoten Islands 171, 216
Lomonosov, Mikhail V. 20
Lomonosov Ridge 20
Long lines 216, **81**
deep sea 216
floating 216, **81**
Lyell, Charles 48–9

Mackerel 214, 219
saury 216
Magnesium 8, 202
Malaysia 218, 220
Management of fisheries 232
Manganese 8, 202
Manganese nodules 184–6
Manuring 224
Marianas Trench 17
Marine Biological Association, Plymouth 173, 206
Marine deposits, ages of 66
definition of kinds of 64–6
distribution of 65
named **20**
Marine environments 64, 67, **35**
life in 67–74, **35**, **61**, **62**, **63**
sounds in 73–4
zones defined in 67–73
Maury, Commodore Matthew Fontaine, U.S.N. 170, **16**

Maximum sustained yield 226
Mediterranean 12, 19, 32, 217, 222–3
Menhaden 208
Mesopelagic zone 68–71, **35**
Meteor Expedition 184
Meteorites 6
Mid-Atlantic Ridge 18–19, 55, 57–8, 172, **30**
Migrations, vertical 212
Milford Haven 222–3, 233
Milkfish 223–4
Minerals in the sea 62, 183–91
  locations of **47**
  mining methods 48, 185
  prospecting methods 193
Mining in the sea, methods of 185–6
Molluscs 203, 212
Molybdenum 8, 202
Moon 4, 80
Monsoons 29, 167, 175, **24**
Mora 231
Moseley, H. N. 169
Muds 212
Murray, Sir John ix, 77, 173
Mussels 210, 224, 228, **91**

Nansen, Fridtjof 204, 210
Nansen-Pettersson water bottle 174, **59**
Naples 173
Neophilina galathea 72
New England 170
New South Wales 233
Newfoundland 168, 209, 216
Newton, Isaac 54
Niger Delta 225
Nitrates 204, 209, 224–5
North Sea 12, 19, 177, 214, 218–19, 227, 233, 235
Nutrient salts 206–10, 224, 235

Ocean, as a global thermostat 34–5
Ocean circulation 26–35
  deep water **28**
  during the ice ages 35–40
Ocean currents 41, 169, 174–5
  deep sea 31–4
  definition of 26–7
  named 27, 34–5, **24**
    Agulhas 29–30
    Benguela 28, 35
    California 227
    Canary 28
    Caribbean 28
    East Greenland 29, 34
    El Nino 35
    Equatorial 209–10
    Gulf Stream 25, *30–1*, 34–5, 39, 168, 170, **27**
    Humboldt *see* Peru
    Kuroshio ('Black Current') 27
    Labrador 29, 34
    North Atlantic (Drift) 168
    North Equatorial 27, 30
    North Pacific 27
    Oyashia 27
    Peru 27, 35, 175, 225, **57**
    Somali 175
    South Equatorial 27, 39
    South Pacific 27
    West Australian 29–30
    West Wind Drift 28
Ocean engineering, definition of term ix
Ocean streams, definition of 26
Ocean, the origin of 1–6
  Greek ideas about 1–2
Oceanography, origin of name viii–ix
  definition of viii
Oceanology, definition of viii
Oceans, bottom sediments in 56, **20**
  life in **35**

Oceans, temperatures in 34–5, 174, **54**
Offtake, permitted, of fish 233
Oil and gas, deposits 175, 187–91
drilling rigs and platforms 187, **45, 46, 49**
Oil pollution *see* Pollution
Organic matter in ocean 203, 222–3, 232–3
Origin of life 1, 4
Otter boards 221
Otter trawl 221
Overfishing and recovery 202, 222–3, 232–3
Overturns 207–8, 210
Oxidation 237
Oxygen, dissolved 205–6, 235
Oysters 203, 210, 228, **91**

Pacific Ocean 9, 11–12, *15–18*, 27–8, 169, 216, 218, 232
Pair trawling 222
Paleomagnetism, origin as a science 53
Pangaea 51
Panthalassa 51
Parasites 224
Pelagic fish 221, **63**
Penguins 236
Peredinians 204
Performance of trawlers 223
Persian Gulf viii, 19–20, 196
Peru 209
Pesticides 214
Philippine Islands 225
Philippine Trench 175
Phoenicians 167
Phosphate, as a nutrient 204, 206, 209, 224–5
Phosphorescence 214
Phosphorus 202
Photosynthesis 198, 203–4, 206–7
Phytoplankton 203, 206–8, 210, 212–13, 224, 235
Piano-wire fish finder 214
Piccard, Jacques 179
Pilchards 208, 211, 214, 219, 232
Pingos 21
Plaice 234
Plankton 201, 208, 228
Plant-eating animals *see* Herbivores
Plate tectonics 60, *passim*, **31**
Plato 74
Plymouth 173, 206, 232
Pogonophores 204
Pollution ix, 202, 235–7, **72–3**
Polynesians 167
Polystyrene, as a pollutant 235
Populations, manipulation of 224, 232
Porcupine Bank 171
Poseidonia 210
Potassium 8, 202
Power blocks 219
Power bulbs 200, **66**
Power fishing 222
Power from the sea 198 *et seq.*
tidal 198
Prawns 212, 214, 216–18, 221, 223–5
Predators 211–12, 216–18, 223–5
Production, gross 205–7, 226
in ponds 224–5
of fish 221, 227
of plankton 203, 226
primary 205–211
regional 206–10, 217
seasonal 206–7, **58**
Proteins 210
Protoplasm 205
Puerto Rico Trench 19
Pumping 225, **70**
Purification, biological 235
Purse seine 81, 215, 219–20

Radioactive material 236
Rainbow trout 226
Rance Estuary, Tidal Power Station 199–200, **66**
Rate of capture 216
Rate of feeding 208
Rays 211, 216
Red Sea 19, 62, 195
Respiration 206
Ridges, ocean 18–21, 30
Ring nets 219, **81**
Romance Deep 19
Rome 235
Rope culture of oysters and mussels 224, **86**
Ross, James 170
Royal Society of London 81, 169, 171
Russia 231

St Croix 194, 225
Salinity 174
Salmon 169, 217–18, 220, 226, 234, **88**
Sand sprays, treated 237, **74**
Santorini *see* Thira
Sardines, Sardinellas 169, 208, 211, 215–16, 219, 229, 232, **64**
Sargasso Sea 25, 209
Sars, G. O. 171
Sauerkraut, as an antiscorbutic 169
Schmidt, Johannes 174
Sciaenid fish 218
Scripps Institution of Oceanography 173, 180
Scurvy 168–9
Sea birds 27–8, 214, 236, **73**
Sea fisheries 173
Sea-floor spread 55, 57, *passim*
Sea slugs 212, **62**
Sea squirts 212
Sea-water, age of 32–3
colour of 209, **56**
distribution of 6–7
estimate of salt content in 9
ionic composition of 9
origin of 1–6, 63
movements of 22–5
salinity of 7–9
some properties of 11
Seabed, the 213–14, 219–20, 228
ownership of 237–9
Seals 228
Seamount *see* Guyot
Seas, the, area covered by 6–7
averages of some metals in 9
regional table of 9
chemical elements in 7–11
table of 8
composition of 7–11
configurations of 11–12
depth statistics in 12
estimate of the water volume contained in 7
principal structural features of (the seabed) defined 12
Seascarps 17
Seaweeds 204, 210, 226
Sediments *see* Deep sea deposits
Semi-submersible drilling platforms 185–7, **46**
Seven seas, the ancient vii
the modern vii
Sewage 235
Shads 211
Sharks 218
Shellfish 210, 223
Ships, surface vessels:
Akademik Knipovitch 228, **65**
Albatross 175
Astragale **45**
Atlantis II 62
Beagle, H.M.S. 75–6, 81, 170, **32**
Buccaneer 32

Ships, surface vessels—*contd:*
Cape Johnson, U.S. 16
Challenger, H.M.S. 170–1, 177, **33**
Dacia **52**
Discovery, H.M.S. 62
Flip (Floating Instrument Platform) 180, **44**
Fram 173
G.O. Sars 229
Galathea 72, 175, 206, 209–10
Glomar Challenger 38, 176–7, 180, 191, **34**
Hirondelle 172
Lightning, H.M.S. 171
Meteor 174
Michael Sars 173
Mikhail Lomonosov 32
Ocean Prospector (platform) **46**
Porcupine, H.M.S. 171
Princesse Alice 172
Retriever **52**
Torrey Canyon 26, 236
Shoals (schools) of fish 219
Shrimp 212, 218, 229
Siebe, Augustus 178
Silversides (*Atherina*) 216
Skates 216
Smith, William 48
Snails 224
Snider-Pellegrini, Antonio 50
Snurrevad *see* Danish seine
Sodium 8, 202
Sole fishery 221, 226, 230, **80**
Sonar, side scanning 191–2, **59**
Sounding 169
Southern Ocean *see* Antarctic Ocean
Spain 222
Sponges 203
Sprats 211, 219–20, 229
Squids 69, 172, 212–13, 215, 230, **61**
jet-propulsion of 213
Starfish 170
Steam trawler 222
Steeman-Nielsen 207
Stocking 224
Stow-net 218, **90**
Stratification of the ocean 169, 194, 209, 216, 225, **54**
Sturgeon 234, **87**
Suakin (Sudan) 195
Submersibles 175, 180, **41**, *see also* Undersea habitats
differences from traditional submarines 180
named, *Archimede* **39**
Beaver **41**
Ben Franklin 31, **27**
Denise 180
Hyco **41**
Manta **41**
Pisces **41**
Shelf Diver **41**
Thresher (U.S. atomic submarine) 179, 182
Trieste 179–80, **39**
Suess, Edward 50
Sugars 205
Sunda Trench 18–19
Sunlight 203–5, 208, 211, 213
Surface waters 207, 209
Surtsey **30**
Sweden 234

Tangle nets 218
Tectonic plates (plate tectonics) 60, *passim*, **31**
Telegraph cables 170
Television, underwater 221
Thira, Island of 74–5
Thomson, Wyville 171
Tidal mills 198, **68**
Tidal power 198, **66, 67**
Tides 4, *40–2*, 80, 169

Tilapia 223
Timber 226
Tonga-Kermadec Trench 17
Trace elements 8, 202
Trade winds 24–5
Transform faults 58–9
Transplantations 234
Traps *see* Fixed fishing gears
Trawls 219, 221, **80**
  diesel and steam compared 223
Trenches (deeps) 12, 15, 17–18, 175, 212, 235, *passim*, **30**
Tropics 209
Tsunami *see* Waves, seismic
Tube-worms 212
Tubers 226
Tunas 216–17, 219, **83**, **84**
Turbot 226, 230
Turbulence 209
Turks 168

Undersea habitats:
  Bacchus **43**
  Conshelf I 182
  Conshelf II 182
  Conshelf III 182, **42**
  FRNS II 179
  Helgoland **43**
  Sealab III 182, **40**
  Seatopia **43**
  Tektite I **43**
Undersea life support systems 181–3
Underwater bulldozer **51**
Unexploited resources 227, **64**
Uniformitarianism 48
United States 219, 223
Upwelling 209, 225, 227, **55**, **56**, **57**, **66**

Vigneron-Dahl Trawl 222
Vikings 167
Vitamins 204
Volcanic island arcs 15, 17, 19, 61
Volcanic islands 75

Waste materials 224, 229
Water, origin of 1–6, 63
  quantity of, contained in hydrosphere 7
  salinity of (sea-water) 7–9
  some properties of 11
Waves, characteristics of wind 42–5
  definition of 42
  fetch of 44
  heights of 42–7
  length of, definition of 42
  seismic (tsunami) 45–7
    warning system for 47
Wegener, Alfred 49, 63, *passim*
West Wind Drifts, definition of 27
Whale-shark 211
Whales 68–9, 172, 175, 208, 211, 213–14, 223, 228, **61**
Whiston, William 50
Winds 170, 174
  definition of 23–5
Winter, plant growth in 208
Wood 203
Worms 203, 210
Wyville-Thomson Ridge 18

Yellowtail 226

Zechstein deposits 188–9
Zinc 8, 202
Zooplankton 68, 203, 208, 210, 212, 227, **35**
  fishery, 228, **65**
Zostera 210